Edexcel International GCSE Biology
Edexcel Certificate in Biology

Revision Guide

Ann Fullick

PEARSON

Published by Pearson Education Limited, a company incorporated in England and Wales, having its registered office at Edinburgh Gate, Harlow, Essex, CM20 2JE. Registered company number: 872828.

www.pearsonglobalschools.com

Edexcel is a registered trademark of Edexcel Limited

Text © Pearson Education Ltd 2011

First published 2011

15 14 13 12 11

IMP 10 9 8 7

ISBN 978 0 435046 76 7

Edited by Patrick Bonham

Proofread by Liz Evans

Original design by Richard Ponsford

Packaged and typeset by Naranco Design and Editorial / Kim Hubbeling

Original illustrations © Pearson Education Ltd 2011

Illustrated by Andriy Yankovskiy

Cover design and title page by Creative Monkey

Cover photo © Pearson Education Ltd: Digital Vision

Printed in Spain by Graficas Estella

Acknowledgements

The author and publisher would like to thank the following individuals and organisations for permission to reproduce photographs:

(Key: b-bottom; c-centre; l-left; r-right; t-top)

Alamy Images: 22cr, 67bc, 78r; iStockphoto: 10tr; Pearson Education Ltd: vi; Science Photo Library Ltd: 67br, 78l

Cover images: Pearson Education Ltd: Digital Vision

All other images © Pearson Education 2011

Every effort has been made to contact copyright holders of material reproduced in this book. Any omissions will be rectified in subsequent printings if notice is given to the publishers.

Websites

The websites used in this book were correct and up to date at the time of publication. It is essential for tutors to preview each website before using it in class so as to ensure that the URL is still accurate, relevant and appropriate. We suggest that tutors bookmark useful websites and consider enabling students to access them through the school/college intranet.

Disclaimer

This material has been published on behalf of Edexcel and offers high-quality support for the delivery of Edexcel qualifications.

This does not mean that the material is essential to achieve any Edexcel qualification, nor does it mean that it is the only suitable material available to support any Edexcel qualification. Edexcel material will not be used verbatim in setting any Edexcel examination or assessment. Any resource lists produced by Edexcel shall include this and other appropriate resources.

Copies of official specifications for all Edexcel qualifications may be found on the Edexcel website: www.edexcel.com

Contents

Doing your best in exams is not always easy. Everyone can do with a little help! Using this book will help you revise everything you need to succeed in International GCSE Biology. We have some tips to help you score the best possible marks in the exam as well.

Using the book

We have put together lots of different features to help you learn what you need to know. They will help you to focus on the really important stuff.

The first few pages and the last few pages look different. They give you handy hints about how to revise and how to take exams successfully. The rest of the book goes through everything you have learned. You will find lots of different ways to learn, and plenty of practice questions too.

Revision advice

Before you get started, this section of the book will give you lots of helpful advice and tips for the best ways to revise. Everyone learns differently, so it is important to discover the revision style that suits you if you want to succeed in your exams. You may need to try out different methods and decide which one works best for you.

Chapter summaries

These contain the main points of each subject that you need to learn. This is the basis of all your revision. Use the chapter summaries and the diagrams and photos to build your own customised revision resource.

Revision questions

You will find a variety of different types of questions that will help you check how well you have revised. They will help you practise for your examination as well.

Exam preparation

The next section of the book looks at preparing for your exams. It gives you a lot of information about what the examiners are looking for when you answer a question. There are some things they are looking for, and others that they really do not want to see. Based on reports from examiners, we will help you work out how to demonstrate your skills and knowledge as well as you possibly can.

How to revise

Revision is key to exam success. So what is revision all about? What are you trying to do when you sit down to revise for your International GCSE Biology exam? There are three main points. You need to:

- **Remember your work:** You want to improve your knowledge and understanding of biology and improve your study skills. You learn the work as you follow your course – but your brain often forgets part of it. When you learn for a test or an exam, you may feel as if you forget everything very quickly! By using different revision techniques, you can help your brain to remember better. Some of these different techniques are described in the next few pages.

- **Organise yourself and your work:** Organise carefully what you have to study. It will make more sense and be easier to learn and remember. Being disorganised makes it harder to learn your work. When your work is organised, you can link ideas together. That makes remembering things easier.

- **Understand the work better:** If you understand your work, you will remember it more easily. It is very hard to learn something you do not understand. So use this book to help you get to grips with everything you have studied. If you find you still do not understand something, go back to your textbook or ask your teacher to explain it again.

Most of your revision will be done on your own, but sometimes working with friends can really help you learn or check out what you know.

Revision tips

Here are some tips to help you revise:

1. Do not leave it too late! Start to revise at least two months before the exams. It is so easy to put off the start of your revision – then as the exams approach it seems overwhelming! It is much better to feel that you have time to take in the work and understand it.

2. Plan your revision carefully. Make sure you leave time to do it all. Making yourself a revision timetable helps, and it can be very rewarding to tick off sections as you complete them.

3. Make your revision active. Simply sitting reading through your textbook or your notes often does not help you learn. Our suggestions in this book will help you find lots of active ways to learn, from making lists to drawing spider diagrams.

4. Split your revision into small chunks. It makes the work much easier to face! And you are more likely to check though a small chunk of revision thoroughly and fully understand it. It is easy to simply end up staring at a book and remember nothing – this is a very poor revision technique.

5. Always make sure you understand your work. If you do not, ask someone to explain it. You cannot learn something you do not understand.

6. Summarise your work as you go through it, then you can use your summary to revise from. There are lots of summaries in this book to get you started.

7. Do not forget to revise your practical work. You may get questions based on practical experiments in your theory papers.

8. Practise answering the exam questions – practice makes perfect! You need to understand exactly what the examiner expects you to do. If you work through all the examples in this book, you will have a good idea of what is expected of you.

Ways of revising

Different people work in different ways. You need to find the way that suits you. However, some ideas are useful for everyone.

Breaking it down

If you look at everything you have to learn in one huge chunk, it is all too easy to give up. To be successful, break your revision down into sections you can manage. There are many ways you can split your work up, so choose one that suits you. For example:

- The order of your class notes – take a lesson or a week at a time.

- The order of the syllabus.

- The order of your textbook. If there are 22 chapters, you could aim to revise about three of them every week.

Nutrition

Human digestive system

Respiratory system

Circulatory system

Skeletal system

Coordination and control

...

Make a list or a table showing how you plan to learn. It is very satisfying to cross off the chunks as you learn them!

Making summaries

Making good summaries based on key words and facts is an important part of any revision programme. You have to really understand something to summarise it well. Your summaries also make a great resource for last-minute revision.

Making connections

Some people find drawing **concept maps** or **spider diagrams** really helpful. Start with a key word or term, such as circulatory system or respiratory system, and then add as many related terms as you can. Link related terms together. Your map or diagram will be unique to the way you learn.

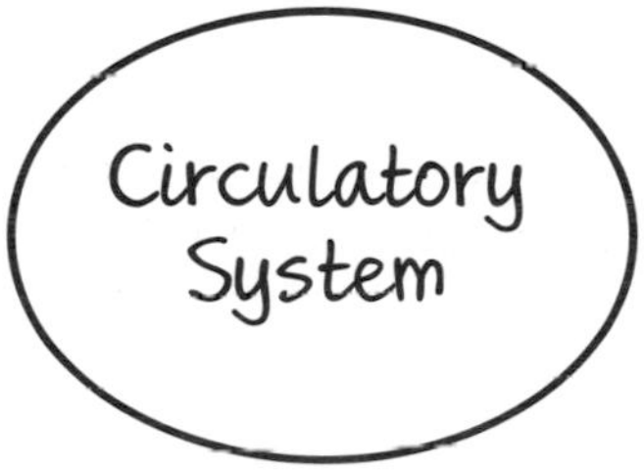

Circulatory system

Need for transport – surface area to volume ratio

Double circulation: heart → lungs → heart → body → heart

Blood vessels – arteries, veins, capillaries

Heart – four chambers, valves, oxygenated and
              deoxygenated blood flow, systole, diastole

Control of heart rate

Blood – functions of plasma, red blood cells,
              white blood cells, platelets, clotting

Lymph and tissue fluid

Heart disease and high blood pressure

For example, if you were building up a map of the topic on the circulatory system, you would start with a central box containing the term 'Circulatory system'.

Circulatory System

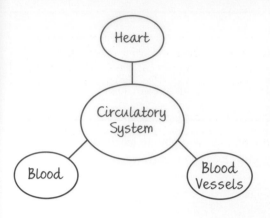

Then you would identify the main aspects of this topic that you need to learn and arrange them in further boxes, such as shown here.

Then break each sub-topic down into smaller units. This helps you to see exactly how the topic you are planning to study fits together. It also makes sure you do not miss any bits out!

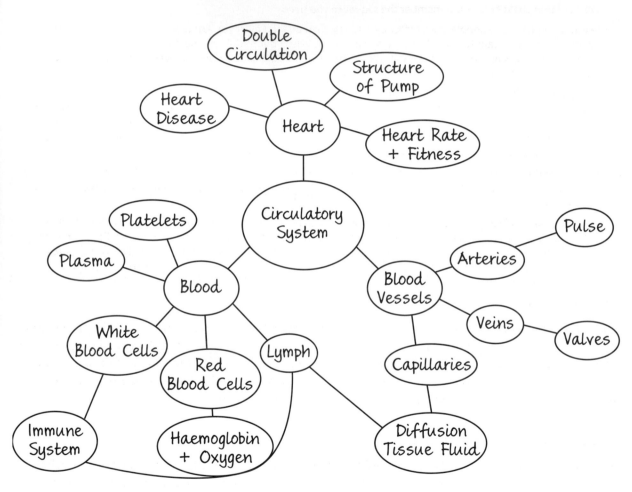

Once you have planned your concept map, put it away. Now do the same process again and see if you put together the same units. Did you miss things out? Did you think of different units you had not thought of the first time? Look at the two maps together and draw yet another version that covers everything you need to study.

Helping your memory

- **Sticky notes:** Do you have difficulty remembering definitions? Biology uses lots of rather technical and scientific words. It can be really helpful to write the words and their definitions on sticky notes or small cards and leave them around your bedroom or on the bathroom mirror so that you see them every day. You can put the words and the definitions on different notes, so you can test yourself!

- **Mnemonics:** To use mnemonics, choose the first letters of the words you want to remember and make up a phrase you can easily remember to remind you. For example, the sequence of events between receiving a nervous stimulus and the response is

  Stimulus → Receptor → Coordination → Effector → Response

  To make a mnemonic out of this, you could remember

  See Really Cool Elephants Roll

  which in turn could help you remember the sequence you need.

- **Reading aloud:** Some people learn better by listening than by reading. If you know that is how you learn best, try recording what you want to learn. Read it aloud into a recorder, your mobile phone, MP3 player or computer. Then you can play it back and listen to the information. Doing this occasionally is good for everyone – it adds a bit of variety and is fun to do.

- **Read, cover, write, check:** This is a technique you have probably used since primary school to help you learn. Now you can apply it to revision. Choose a short unit of work. Read it through carefully several times. Now cover up the topic and write down as many of the main points as you can remember. Now check what you have written against the original card, adding anything you have forgotten in a different colour. You can use this card when you revisit the work. Your mind will focus harder on the words in a different colour – the words you forgot first time around!

- **Working with a friend:** Spending time with a friend can be a very effective way of revising. You can quiz each other and read bits out to each other. There are lots of ways you can work together to make learning more effective – and more enjoyable.

- **Flow diagrams:** When you need to learn something in sequence, it can be really helpful to draw a flow chart or spider diagram of what you are trying to learn. Study your diagram, then cover it up and redraw it. Compare the new version with the first diagram. Mark the parts you did not remember well in a different colour. Repeat the process later. Your mind will focus harder on the words marked in a different colour when you come to learn it again.

Practice questions

One of the most important aspects of revision is to try out practice questions once you have learned a topic. This is helpful for two reasons. It helps you check that you really have learned the biology. It also gives you a lot of practice at answering questions before you have to do it in the exam. Appendix A, at the end of this book, gives you all sorts of hints about how to answer questions well and what the examiners are looking for. Take time to read that section *before* you try to answer any questions.

Organising your revision

To revise successfully, you need to be organised. Flicking though your textbook while watching TV will not help you prepare for your exams! Here are a few key ways to make your revision as effective as possible.

- **Get the timing right.** You cannot revise for hours at a time and hope to actually remember anything. Most people find it hard to concentrate for more than 20 to 30 minutes at a time. If that is what suits you, work and then take a short break. Grab a drink, watch TV for 5 minutes – do something that relaxes you – before returning to your studying. After about an hour and a half – three or four short sessions – have a longer break with a walk or a chat before returning to work. You will revise much more effectively if you take a proper break. In fact, for most people an hour and a half of one subject is quite enough for the day, and it might be better then to do some revision for a different exam.

Some people prefer to immerse themselves in their work for a longer period between breaks. If it suits you to focus on your biology revision for 45 minutes or an hour at a time, that is fine. But remember, everyone needs breaks, and a fresh brain is an effective brain!

Do not spend the night before the exam frantically trying to read through your whole course. At most, have a quick look through your summary cards. Then relax and get a good night's sleep, so you can do your best and show the examiners what you know.

- **Find the right place.** The best place to revise is somewhere quiet on your own, so you can work without being disturbed. If you cannot find a private space, try to work somewhere as quiet as possible. Watching TV at the same time is not a good idea! Some people find listening to music helps them focus and revise, while others do not – just do what feels right for you. But you cannot have music in the exam room, so do not rely on it to help you remember things.

There is some research which shows that linking a new scent to a particular subject can help trigger your memory. So you could try using a particular scent, aftershave or essential oil every time you revise biology, then wash it off afterwards. Remember to put some on the day of the exam. It might help – and it will not do any harm!

- **Remember to relax.** It is important not to work all the time. You need to take some rest and relaxation to keep your body healthy and your mind fresh. Take some time out with friends and family when you can forget about revision and exams. Then when you next sit down to study, you will come to it enthusiastically and ready to learn as effectively as possible.

Good luck!

Chapter 1: Life processes

The characteristics of living things

Most living organisms have eight life processes in common:

- **nutrition** (needing food)
- **excretion** (removing waste products)
- **movement** (of all or part of the organism)
- **growth and development**
- **respiration** (getting energy from food)
- **response to stimuli** (changes in the surroundings)
- **reproduction** (producing offspring)
- **controlling their internal environment.**

Cell structures

- All living organisms are based on units known as **cells**. We usually draw unspecialised cells to show features common to most cells.
- There are eight life processes that are common to all living organisms.
- Unspecialised animal cells all have a **cell membrane**, **cytoplasm**, **nucleus,** and **mitochondria**.
- Unspecialised plant cells also have a **cellulose cell wall** and may have a **permanent vacuole**. Cells from the green parts of a plant contain **chloroplasts**, which contain **chlorophyll** for **photosynthesis**.
- **Enzymes** act as **biological catalysts** in metabolic reactions. They speed reactions up and control them.
- The functioning of enzymes can be affected by temperature and pH.
- Cells respire **aerobically** and **anaerobically** to release energy from food.
- Substances can move into and out of cells by **diffusion**, by **osmosis** and by **active transport**.
- In multicellular organisms the cells of the embryo **differentiate** to form **specialised cells** which carry out particular functions in the body. The levels of organisation in an organism are **organelles**, **cells**, **tissues**, **organs** and **organ systems**.

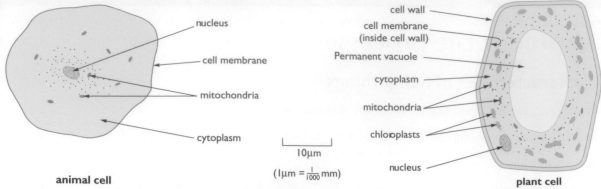

Figure 1.1 *Animal and plant cells as seen under the light microscope*

animal cell

plant cell

Enzymes

Enzymes are **biological catalysts** that control all the reactions in a cell. A catalyst speeds up a reaction without being used up or changing the reaction. Enzymes are proteins. The shape of the protein molecule makes an active site. The substrate attaches to this active site. The reaction takes place and products are formed. Anything that changes the shape of the active site stops the enzyme working (**denatures** the enzyme), e.g. heat or pH.

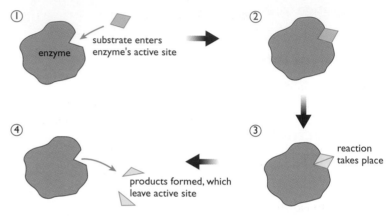

Figure 1.2 *The lock-and-key mechanism of how enzymes work*

Experimental evidence

You can investigate the effect of temperature on the enzyme amylase using starch and iodine, putting the mixture in water baths at different temperatures.

Cellular respiration

Cells break down small food molecules that have been assimilated from the gut to release the stored chemical energy in the process of **cellular respiration**. The chemical energy is used for muscle contraction, active transport, building up large molecules and cell division.

In **aerobic respiration** oxygen is used to oxidise food. Carbon dioxide and water are released as waste products:

Glucose + oxygen → carbon dioxide + water (+ energy)

$C_6H_{12}O_6$ + $6O_2$ → $6CO_2$ + $6H_2O$ (+ energy)

Sometimes cells have to respire with no oxygen available. This is **anaerobic respiration**. Glucose is not completely broken down and less energy is released. In yeast cells the waste products of anaerobic respiration are **ethanol** and carbon dioxide. In animal cells the waste product is **lactic acid**.

Glucose → ethanol + carbon dioxide   (+ some energy)

Glucose → lactic acid   (+ some energy)

Anaerobic respiration takes place in muscle cells when they are working hard during exercise. Once exercise stops, extra oxygen is still needed to break down the lactic acid fully. The oxygen needed is known as the oxygen debt.

Experimental evidence

You can test for carbon dioxide production during respiration using limewater.

You can test for heat production during respiration using germinating peas in vacuum flasks.

Movement into and out of cells

Diffusion is the net overall movement of particles from an area of high concentration to an area of lower concentration. It is caused by the random movement of particles in gases and liquids. Diffusion is **passive** – it takes place *down* a concentration gradient and does not use energy.

Diffusion rates are affected by **concentration**, **temperature** and available **surface area**.

Osmosis is a special type of diffusion where only water moves across a partially permeable membrane, from an area of high concentration of water to an area of lower concentration of water *down* the concentration gradient.

Cell membranes are partially permeable, so osmosis takes place across cell membranes.

In active transport, substances are moved against a concentration gradient or across a selectively permeable membrane. Active transport uses energy produced by cellular respiration.

Experimental evidence

You can demonstrate diffusion using agar blocks dyed with potassium permanganate.

You can demonstrate osmosis using membrane bags, by using potato tissue or by observing plant cells.

Specialised cells

In multicellular organisms the cells become specialised for a particular function in the body, e.g. sperm cells, nerve cells, palisade cells. They are adapted to carry out a particular job. Specialised cells group together to form a **tissue**. A tissue carries out a particular function, e.g. muscle tissue contracts. Different tissues work together, forming an **organ**. Organs carry out functions in the body, e.g. the heart pumps blood. An **organ system** involves several different organs working together to carry out particular functions in the body, e.g. the digestive system or the circulatory system.

The features of a specialised cell show how it is adapted for its function, e.g. the number of mitochondria tell you how much energy the cell uses.

Figure 1.3 *The relationship between a cell and the whole body*

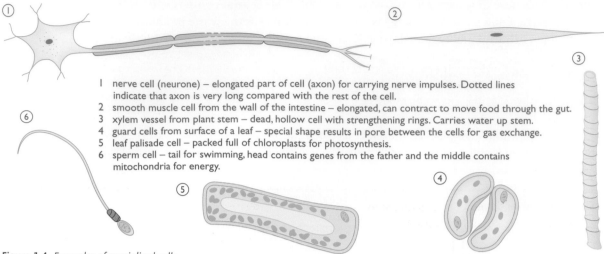

I nerve cell (neurone) – elongated part of cell (axon) for carrying nerve impulses. Dotted lines indicate that axon is very long compared with the rest of the cell.
2 smooth muscle cell from the wall of the intestine – elongated, can contract to move food through the gut.
3 xylem vessel from plant stem – dead, hollow cell with strengthening rings. Carries water up stem.
4 guard cells from surface of a leaf – special shape results in pore between the cells for gas exchange.
5 leaf palisade cell – packed full of chloroplasts for photosynthesis.
6 sperm cell – tail for swimming, head contains genes from the father and the middle contains mitochondria for energy.

Figure 1.4 *Examples of specialised cells*

Questions

1 Copy and complete these sentences. Use the words below to fill in the gaps.

excrete	**move**	**internal environment**
reproduce	**respiration**	**stimuli**	**eight**

There are _____ life processes common to all living organisms. They all need food and release energy from their food by _____. All living organisms _____ to get rid of waste and _____ all or part of their body. When organisms _____ they make more of themselves, and these offspring grow to adult size. Living organisms respond to _____ and can control their own _____.

2 a) Draw and label an unspecialised animal cell.

b) Explain the function of each of the structures you have drawn and labelled.

3 a) List all the main structures you would expect to find in a plant cell taken from the leaf of a plant.

b) Three of these structures would not be found in an animal cell. Explain the function of these three plant cell features.

c) How would a plant root cell be different from a plant leaf cell? Explain why it would be different.

4 Some students investigated the breakdown of starch using the enzyme amylase (found in the saliva in your mouth). In one test tube they kept starch solution at room temperature. In two other tubes they mixed the starch solution with the enzyme amylase, and then placed one tube in a water bath at body temperature. They took samples from each tube every minute and mixed them with iodine on a spotting tile. Iodine turns blue-black in the presence of starch. The results are shown below.

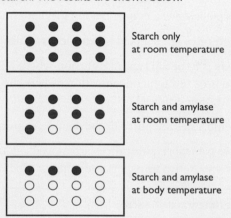

Starch only
at room temperature

Starch and amylase
at room temperature

Starch and amylase
at body temperature

a) What effect does amylase have on starch and what is your evidence for this?

b) What do the results tell you about the effect of temperature on the action of the enzyme amylase?

c) Why is one tube of starch solution kept at room temperature without the addition of the enzyme?

d) What do you predict would happen to the activity of the enzyme if acid from the stomach was added to the mixture, and why?

5 When you move around normally, you produce the energy your body needs by **aerobic respiration**. If you are exercising hard and there is not enough oxygen reaching your muscles, your body gets its energy using **anaerobic respiration**.

a) Give a word equation for aerobic respiration.

b) What is produced when animal cells respire anaerobically?

c) What are the main differences between aerobic and anaerobic respiration?

6 *a)* Why do sharks find an injured fish – or person – so easily?

b) What is meant by the net movement of particles?

c) What factors most affect the rate of diffusion in a liquid or a gas?

7 The figure below shows an experiment to demonstrate osmosis taking place.

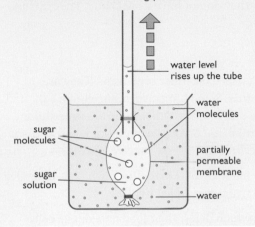

water level
rises up the tube

water molecules

sugar molecules

partially permeable membrane

sugar solution

water

a) Explain what is happening.

b) Explain what would happen if you set up a similar experiment using a partially permeable bag containing pure water in a beaker containing strong sugar solution.

c) How does osmosis differ from diffusion?

8 *a)* What are specialised cells?

b) How can the number of mitochondria in a cell help you to decide its function?

c) Define the following terms and for each one give an example:
   i) tissue
   ii) organ
   iii) organ system

d) Draw a simple diagram to explain how specialised cells are related to organ systems in the body.

Chapter 2: The variety of living organisms

- Plants are multicellular and carry out photosynthesis. Examples: maize, peas, ferns.

- Animals are multicellular. All animals need to feed on other living organisms to get their energy. Examples: humans, insects, worms.

- Fungi can be multicellular or unicellular. They absorb food from other living organisms. Examples: moulds, yeast.

- Protoctists are mainly microscopic, single-celled organisms. Examples: protozoa, algae.

- Bacteria (the prokaryotes) are microscopic, single-celled organisms that are much smaller than protoctists. Examples: *Lactobacillus*, *Pneumococcus*.

- Viruses are even smaller than bacteria. They are parasites that can only reproduce inside other living cells.

- Bacteria, viruses, protoctists and fungi can all act as pathogens, causing disease in other organisms.

Classification

There are around 10 million species of animals and plants. Biologists classify them into groups based on similarities in their structure and functions.

The major groups are plants, animals, fungi, protoctists, bacteria and viruses.

Plants

Plants are all multicellular. They carry out photosynthesis using carbon dioxide and water to make sugar and oxygen. They capture energy from the Sun using chlorophyll in their chloroplasts. They store carbohydrates as starch. Plant cell walls are made of cellulose. There are flowering plants such as cereals and grasses as well as peas and sunflowers. There are simpler non-flowering plants such as mosses and ferns.

Animals

There is a huge variety of animals. They are all multicellular organisms which cannot carry out photosynthesis as they never contain chlorophyll. They eat other organisms to get their energy. Animal cells do not have a cell wall. Most animals have a form of nervous system for coordination. They move their whole bodies about. The carbohydrate stored in animal cells is **glycogen**.

Vertebrates are animals that have a backbone (vertebral column). They include fish, amphibians, reptiles, birds and mammals. All other animals are **invertebrates** – they do not have a backbone. They include sponges, molluscs, worms, starfish, insects and crustaceans.

Fungi

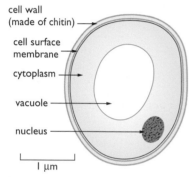

Figure 2.1 *A yeast cell*

Fungi are organisms that feed off other organisms. They are **saprophytes** or **parasites**. Some fungi are multicellular, e.g. mushrooms, toadstools and moulds. Other fungi are unicellular. Fungal cells have cell walls that are made of chitin, whereas plant cell walls are made of cellulose.

Multicellular fungi are made up of a **mycelium**, which is a tangled network of thread-like structures called **hyphae**.

Fungi secrete enzymes which digest their food outside the cells. They then absorb the digested products. This is saprophytic nutrition. Many fungi act as decomposers.

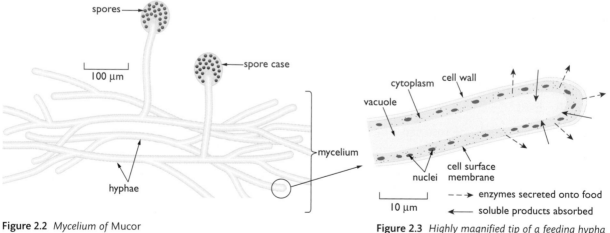

Figure 2.2 *Mycelium of* Mucor

Figure 2.3 *Highly magnified tip of a feeding hypha*

Protoctists

Some protoctists make their own food by photosynthesis (**algae**) and some feed on other organisms (**protozoa**). Most are unicellular but some are multicellular. Some cause disease, e.g. *Plasmodium* causes malaria. Many protoctists are very small (microscopic) but some are very large, e.g. some seaweeds.

Bacteria (Prokaryotes)

These are tiny, single-celled living organisms that are much smaller than animal cells. Common shapes include spheres, rods and spirals. Bacteria have a different structure from that of animal and plant cells. They have a complex cell wall made up of polysaccharides and proteins, different from the cellulose of plant cell walls and the chitin of fungal cell walls. Bacteria do not have a nucleus. The genetic material forms a circular loop in the cytoplasm. Small extra circles of DNA are known as **plasmids**. Some bacteria contain chlorophyll and can photosynthesise. Most feed on dead or other living organisms. Some bacteria have slime capsules for protection or flagella for movement.

Prokaryotes

Prokaryotes are single-celled organisms that do not have a nucleus. Most prokaryotes are bacteria but they also include the *archaea*. They are similar to bacteria but some of their cell chemistry is different.

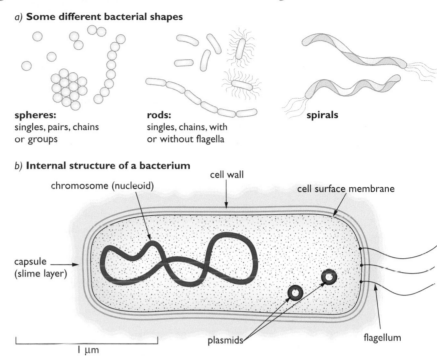

a) **Some different bacterial shapes**

spheres: singles, pairs, chains or groups

rods: singles, chains, with or without flagella

spirals

b) **Internal structure of a bacterium**

cell wall

chromosome (nucleoid)

cell surface membrane

capsule (slime layer)

plasmids

flagellum

1 μm

Figure 2.4 *The shape and structure of bacteria*

Viruses

All viruses are parasites and can only reproduce inside the living cells of a host. Viruses are even smaller than bacteria. Virus particles are very simple and are found in many different geometrical shapes. They have a core of genetic material surrounded by a protein coat. Viruses contain either RNA or DNA but not both. All natural viruses cause disease, and every type of living organism, even bacteria, can be infected by viruses. Each type of virus causes one specific disease in the particular organism it is adapted to infect, e.g. flu, HIV/AIDS.

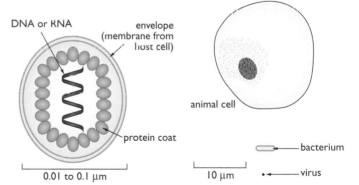

DNA or RNA

envelope (membrane from host cell)

protein coat

0.01 to 0.1 μm

Figure 2.5 *The structure of a virus*

animal cell

bacterium

virus

10 μm

Figure 2.6 *The relative sizes of an animal cell, a bacterium and a virus. Plant cells are bigger than animal cells*

1 Classification divides all living organisms into groups. Copy the table below and complete the empty boxes. Write **yes**, **no** or **sometimes** to indicate whether the organism is multicellular, give an example of an organism of this type, and describe how organisms of this type feed.

Type of organism	Example	Way(s) of feeding	Multicellular?
animal	cow	eats other animals/ plants	Yes
plant			
protoctist			
bacterium (prokaryote)			
virus			

2 For many years people thought there were only two kingdoms in the living world – plants and animals. Although we now know there are other kingdoms, plants and animals are still very important.

a) Copy and complete this table to show the main characteristics of plants and animals.

Plants	Animals
	Multicellular
main carbohydrate storage compound is starch	

b) Give a difference and a similarity between plants and fungi.

3 a) What does classification mean?

b) What do scientists look for when classifying organisms?

c) What determines if an animal is called a vertebrate or an invertebrate?

4 a) Name two different types of fungi.

b) How do fungi differ from plants?

c) Explain the following terms:

i) hyphae

ii) mycelium

iii) spore

5 a) Draw and label a yeast cell.

b) Draw and label one hypha of a mould such as *Mucor*.

c) A fungus such as *Mucor* has a special way of obtaining its food.

i) Describe how *Mucor* obtains its food.

ii) What is this method of feeding called?

6 Bacteria and viruses are very small. Bacteria have a structure a bit like a plant cell with a cell wall, but their genetic material is not in a nucleus. Viruses are even smaller than bacteria and are made up of a simple protein coat containing a small amount of genetic material.

Copy these diagrams of a bacterium and a virus and use the information above to help you label your diagrams.

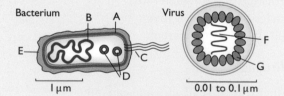

7 Bacteria and viruses are both microorganisms, but they are very different. Explain the differences between bacteria and viruses in terms of:

a) their size

b) the way they carry out the processes of living organisms

c) their impact on human life.

Chapter 3: Breathing and gas exchange

- The gas exchange system takes air into and out of the body to supply oxygen and remove carbon dioxide.
- The movement of air in and out of the lungs is brought about by breathing movements of the ribs and diaphragm. This air movement maintains a steep concentration gradient for the diffusion of oxygen and carbon dioxide between the blood and the air in the lungs.
- The complete gas exchange system includes the ribs, the intercostal muscles, the diaphragm, the trachea, bronchi, bronchioles, alveoli and pleural membranes of the thorax.
- Gas exchange takes place in the lungs and depends on efficient diffusion of the gases.
- The alveoli are adapted for efficient gas exchange.
- Cigarette smoke contains nicotine, carbon monoxide, tar and other chemicals.
- There is strong evidence for a link between smoking and diseases of both the lungs and the circulatory system.

Breathing

Cellular respiration is the process that releases energy from food in your cells. Your cells need a constant supply of oxygen for this to happen aerobically. Your cells also need to have the waste carbon dioxide produced removed.

Your gas exchange (respiratory) system is an organ system adapted to allow the exchange of oxygen and carbon dioxide to take place as efficiently as possible. You bring air into your body through your mouth and nose. It is made warm and moist on the way in, and dirt and pathogens are filtered out of it by cilia and mucus. The lungs are the organs where gas exchange takes place. They are specially adapted to make the exchange of oxygen and carbon dioxide as efficient as possible.

Structure of the human gas exchange system

nasal passages	warm, clean and add moisture to the air
epiglottis	stops food getting into lungs when you swallow
oesophagus	carries food to stomach
larynx	voice box
trachea	tube with incomplete rings of cartilage carries air to lungs; lined with cells making mucus, and cells with cilia which move the mucus away from the lungs
left bronchus	carries air to lung
bronchioles	carry air to lungs
alveoli	tiny air sacs adapted for gaseous exchange
diaphragm	sheet of muscle with a fibrous middle part which is domed; it helps make breathing movements and separates the thorax from the abdomen
ribs	bones that protect and ventilate lungs
internal intercostal muscles	pull ribs down and in when you breathe out
external intercostal muscles	pull ribs up and out when you breathe in
pleural membranes	thin moist membranes forming an airtight seal around lungs and separating inside of thorax from lungs
pleural fluid	liquid filling pleural cavity and acting as lubrication, so surfaces of lungs do not stick to inside of chest wall

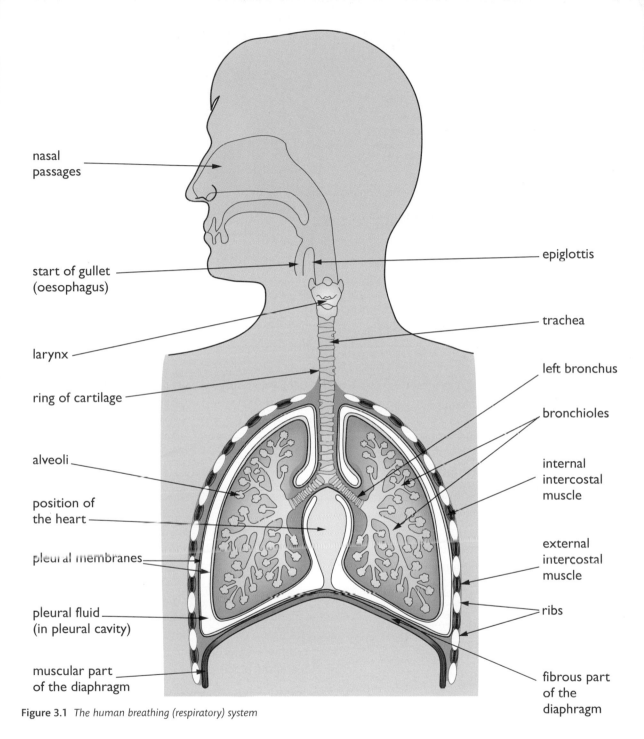

nasal passages

epiglottis

start of gullet (oesophagus)

trachea

larynx

left bronchus

ring of cartilage

bronchioles

alveoli

internal intercostal muscle

position of the heart

pleural membranes

external intercostal muscle

pleural fluid (in pleural cavity)

ribs

muscular part of the diaphragm

fibrous part of the diaphragm

Figure 3.1 *The human breathing (respiratory) system*

Ventilating the lungs

Air is moved into and out of the lungs by movements of the ribs and diaphragm, as shown in figure 3.2. Ventilation brings in oxygen-rich air and removes air containing carbon dioxide. It maintains steep diffusion gradients in the alveoli. There is always more oxygen and less carbon dioxide in the air in the lungs than in the blood. The ventilation movements of the ribs and diaphragm bring about changes in the volume and therefore the pressure of the chest cavity.

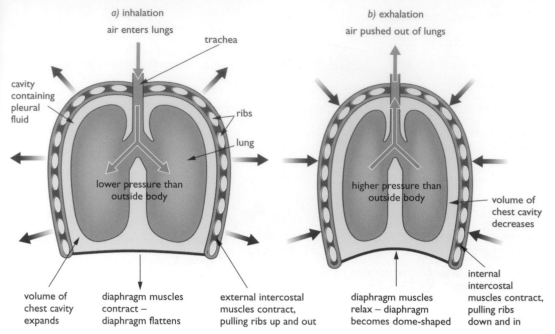

Figure 3.2 *Volume and pressure changes in the chest during breathing movements move air into and out of the lungs*

Experimental evidence

You can investigate breathing movements by resting your hands on your ribcage as you breathe in and out.

You can investigate the effect of exercise on breathing rates by measuring breathing rates at rest and after different periods of exercise.

Gas exchange in the alveoli

The lungs contain millions of tiny air sacs called alveoli. They are adapted for efficient gas exchange. Blood is pumped from the heart to the lungs and passes through the network of capillaries surrounding the alveoli. Carbon dioxide diffuses from the blood into the air in the alveoli. Oxygen diffuses from the air in the lungs into the blood. The oxygenated blood travels back to the heart to be pumped around the body. Oxygen dissolves in the layer of fluid lining the alveolus before it diffuses into the blood. Alveoli have:

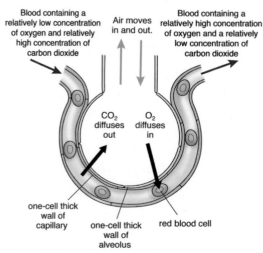

Figure 3.3 *Gas exchange in an alveolus*

- a large surface area
- a rich blood supply, which removes oxygen from and delivers carbon dioxide to the air in the alveoli, maintaining a steep diffusion gradient between the alveoli and the blood
- short diffusion distances between the air and the blood.

Smoking and health

Nicotine is the addictive drug found in tobacco. It is addiction to nicotine that makes it very hard to give up smoking. Carbon monoxide reduces the amount of oxygen carried in the blood of smokers. Tar and other chemicals in tobacco smoke can cause lung cancer, bronchitis, emphysema, and diseases of the heart and blood vessels.

Tobacco smoke has the following effects.

- Cilia are destroyed so dirt and bacteria are not removed.

- Emphysema – the walls of the alveoli are damaged and break down to form large irregular air spaces which do not exchange gases efficiently.

- Lung cancer – tar and other chemicals cause cells to mutate and form cancers in the lungs and throat.

- Carbon monoxide binds to haemoglobin, lowering the oxygen levels in the blood. In pregnant women, this deprives the fetus of oxygen and can lead to smaller babies and stillbirths.

- Smoking also affects the circulatory system and causes an increased risk of heart attacks and strokes.

Questions

1 Copy and complete these sentences. Use the words below to fill in the gaps.

**respiratory system   breathing structures
carbon dioxide   lungs   air passages**

The job of your _____ is to get fresh supplies of air containing oxygen into your _____, and to get rid of waste_____ produced by your body. The breathing system has three main parts: the _____, the alveoli and the

_____.

2 Look carefully at the words and definitions. Match each word to its correct definition and then copy them out.

alveoli	the smallest air passages in the lungs
trachea	the upper part of the body containing the lungs
lungs	the main air passage leading in from the mouth and nose
diaphragm	the body organs where gas exchange takes place
bronchioles	millions of tiny air sacs making up the gas exchange tissue
thorax	large sheet of muscle separating thorax from abdomen

3 For gas exchange in the lungs to work effectively, we need to move air in and out of the lungs regularly. We do this by breathing or ventilation. Our breathing movements involve the muscles between the ribs, and the muscles of the diaphragm. Explain carefully, using diagrams if you feel they will help, the events that take place:

a) when you breathe in

b) when you breathe out.

4 The table above shows the effect of exercise on the breathing rate of three people.

Activity	Number of breaths taken per minute		
	Person A	Person B	Person C
Rest	21	15	18
20 step-ups per minute	29	21	25
50 step-ups per minute	40	30	34

a) Plot a bar chart of these results to make it easier to compare them.

b) Which person do you think is the fittest of the three? Which do you think is the least fit? Explain your answers.

c) What else would happen to the breathing as well as the rate going up?

d) Why does our breathing change when we exercise?

5 Copy this diagram and match the labels below to the statements on the next page.

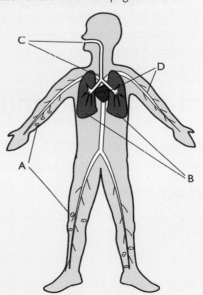

1 You breathe to take air in and out of the body.

2 In your lungs you exchange waste carbon dioxide from the blood for oxygen from the air.

3 Your circulatory system makes sure every cell in the body gets the oxygen it needs.

4 In the cells of your body, respiration uses oxygen from the air to release the energy from your food, making carbon dioxide and water as waste products.

6 The air you breathe in contains about 20% oxygen and only 0.04% carbon dioxide. The air you breathe out contains about 16% oxygen and 4% carbon dioxide. What happens in your lungs to bring about this change?

7 Here is a diagram of a single alveolus from the lungs. The alveoli are very specialised structures to allow the exchange of gases in the lungs.

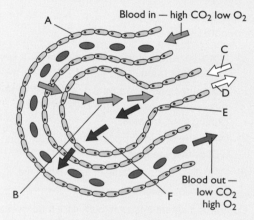

Blood in — high CO_2 low O_2

A
C
D
E
B
F

Blood out — low CO_2 high O_2

a) Copy and label this diagram.

b) Describe three features of alveoli which make this exchange as effective as possible.

8 Smoking is bad for the health.

a) Name three components of tobacco smoke.

b) Describe three ways in which tobacco smoke damages the lungs or other body systems.

c) Mrs Brown is a heavy smoker. She has one baby which is very small when it is born. Her next baby is born dead.

i) Why are smokers more likely to have very small babies than non-smoking mothers?

ii) If a mother smokes, how can it affect the baby once it is born?

9 Here are descriptions of three different diseases that affect the breathing system. Use your knowledge of how the breathing system works to explain why these diseases make breathing so difficult and result in a lack of oxygen in the body.

a) In cystic fibrosis the lung cells are covered by a very thick, sticky mucus which fills the alveoli and blocks the bronchioles. It makes lung infections more likely and the lungs have to be cleared of mucus by physiotherapy at least twice a day. People with cystic fibrosis are often short of breath, particularly just before they have physiotherapy.

b) COPD (emphysema) is a disease often caused by smoking where the structure of the alveoli breaks down, resulting in lungs with much larger air sacs than normal. These large spaces may fill with fluid. People with COPD are always short of breath, and as the disease gets worse they may need to breathe pure oxygen.

c) In asthma the linings of the air passages swell and produce extra mucus. During an asthma attack, people find it very hard to breathe and the air is forced in and out of their chests with a wheezing sound.

Chapter 4: Food and digestion

- A balanced diet should include carbohydrate, protein, lipid, vitamins, minerals, water and dietary fibre.
- There are simple chemical tests for glucose and starch.
- Energy requirements vary with activity levels, age and pregnancy.
- The human alimentary canal includes the mouth, oesophagus, stomach, small intestine, large intestine and pancreas.
- The process of breaking down food involves ingestion, digestion, absorption, assimilation and egestion.
- Food is moved through your gut by peristalsis.
- Food is broken down by enzymes.

- Bile is released to neutralise stomach acid and emulsify fats.
- The villi of the small intestine increase the surface area available for the absorption of digested food molecules.

The food you eat is used in three ways: it provides energy for your cells, it provides the material needed for growth and repair, and it provides the resources you need to fight disease and stay healthy. You need to eat a balanced diet to stay healthy.

Carbohydrates are sugars and starches. They are made up of atoms of carbon, hydrogen and oxygen joined together. Carbohydrates are found in bread, pasta and rice. They provide energy for the cells. Simple carbohydrates like glucose and sucrose are sugars. Complex carbohydrates like starch, glycogen and cellulose are long chains of sugars such as glucose or fructose joined together.

Experimental evidence

Reddish-brown iodine solution turns blue-black with starch.

Blue Benedict's solution turns orange-red when heated with simple sugars like glucose.

You can determine the energy content of a food sample by burning it and using it to heat a measured sample of water. This is a simple calorimeter.

Proteins are long chains of amino acids found in foods such as meat, fish and cheese. Proteins are used for body building. They contain carbon, hydrogen and oxygen but they have nitrogen and sulfur atoms as well.

Lipids are made up of fatty acids and glycerol joined together. They are found in foods such as butter, corn oil and eggs. They contain a lot of energy and are used as an energy store in the body. Lipids are made up of carbon, hydrogen and oxygen.

EXAMINER'S TIP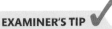

Remember, carbohydrates and lipids contain only carbon, hydrogen and oxygen. Proteins also contain nitrogen and sulfur.

Minerals and vitamins

Minerals and vitamins are needed in tiny amounts in your diet. If you do not get enough of them, you will suffer deficiency diseases. Examples are shown in the table.

Mineral or vitamin	Role in the body	Examples of foods that supply it
A	makes chemicals in retina, protects surface of eye and connective tissue	fish liver oils, liver, butter, margarine, carrots
C	sticks together cells lining surfaces in the body, e.g. mouth	fresh fruits (especially citrus fruits) and vegetables
D	helps bones absorb calcium and phosphorus	fish liver oils, cream, butter, also made in skin in sunlight
calcium	makes bones and teeth	dairy products, fish, bread, vegetables
iron	makes haemoglobin in red blood cells to carry oxygen	red meat, liver, eggs, green leafy vegetables

Water is important in your body as a solvent, for breaking up large molecules by hydrolysis, and for carrying substances around your body. You should drink plenty of clean water every day.

You need **fibre** in your diet for bulk to give the muscles of the gut something to work on. It also absorbs lots of water.

Energy requirements

Different foods contain different amounts of energy. Energy is measured in joules (calories). It is important to take in the right amount of energy for your body's needs. If you take in too much energy, you will get fat. With too little energy from your food, you will become very thin. The amount of energy you need varies with your age, with how active you are, and whether you are pregnant.

Digesting your food

The food you eat is made up of large insoluble molecules. Your cells need small soluble molecules. The physical and chemical breakdown of your food by enzymes takes place in your digestive system.

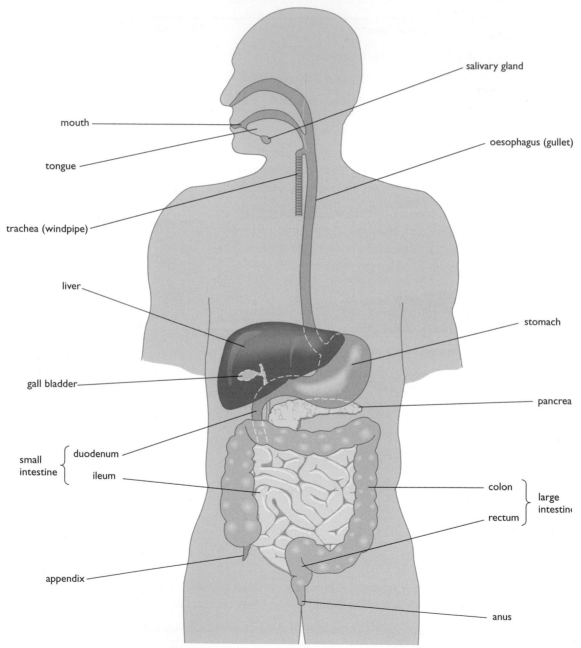

Figure 4.1 *The human digestive system*

The alimentary canal (gut) is a muscular tube that runs from your mouth to your anus. Food moves through your gut by waves of muscle contractions. You produce enzymes which digest food in glands that are found in several different organs of your digestive system. These include your salivary glands, your pancreas and your ileum. Bile is made by your liver and stored in your gall bladder until it is needed. It neutralises stomach acid and emulsifies fats, making them easier for lipase to digest.

Type of enzyme	Where found in gut	What does it act on?	What are the breakdown products?
carbohydrase, e.g. amylase, maltase	salivary glands, pancreas, small intestine	starch	glucose
protease, e.g. pepsin, trypsin	stomach, pancreas, small intestine	protein	amino acids
lipase	pancreas, small intestine	lipids (fats and oils)	fatty acids and glycerol

Experimental evidence

You can investigate the effect on activity of gut enzymes such as amylase and pepsin when the temperature or pH of the environment is changed.

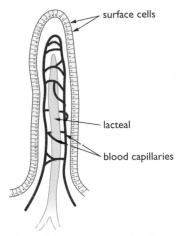

surface cells

lacteal

blood capillaries

Figure 4.2 *The structure of a villus*

Absorption: The products of digestion are absorbed into your bloodstream in the small intestine. The lining of the small intestine is covered in villi, which are adapted for effective diffusion. They have a large surface area, a good blood supply to maintain steep concentration gradient, and short diffusion distances.

Assimilation: The digested food products are absorbed into the cells of the body by diffusion and assimilated. They are used for energy and to build up into chemicals needed by the cells.

Egestion: Water, undigested food, enzymes, dead cells, bile pigments and mucus move through the large intestine. Water is removed. The remaining material is passed out of the anus as faeces.

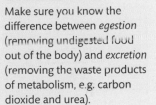

EXAMINER'S TIP

Make sure you know the difference between *egestion* (removing undigested food out of the body) and *excretion* (removing the waste products of metabolism, e.g. carbon dioxide and urea).

Questions

1 Copy and complete these sentences. Use the words below to fill in the gaps.

carbohydrates	malnutrition	healthy
fats	energy	chemicals

It is important to eat the right amounts of food to remain fit and _____. Food provides you with _____ and the different _____ you need to keep your body working properly. The main types of food are _____, proteins and _____. Eating too much, too little or the wrong sort of food can result in _____ .

2 Copy and complete each sentence, choosing the correct ending from the list on the next page.
   a) Carbohydrates are found in foods such as
   b) Both carbohydrate and fats supply energy
   c) Fats are found in foods such as
   d) Too much, too little or the wrong sort of food
   e) Proteins, important for growth and replacing cells,

Choose endings from:
- cheese, butter and margarine.
- are found in meat, fish, eggs and pulses.
- cereals, fruits and root vegetables.
- but the energy in the carbohydrates can be used more easily by the body.
- causes malnutrition.

3 a) Name two minerals that are needed in the diet, and explain what they are used for in the body.

b) Name one fat-soluble vitamin and one water-soluble vitamin, and explain what they are used for in the body.

c) Mrs Lynch enjoys food that is rich in fat and carbohydrate. She spends most of her time at home and doesn't walk far. State what diet-related health problem she is most likely to suffer from.

d) How could Mrs Lynch improve her diet?

4 People who eat a diet rich in fibre generally produce a lot of faeces, and food passes through their gut in about 24 hours. People who eat a low-fibre diet pass much smaller amounts of faeces and food can stay in their gut for several days. Some evidence links the amount of fibre in the diet with the risk of developing cancer of the bowel. Other evidence does not make this link.

Number of men aged 36–64 suffering bowel cancer per year (per 100 000)

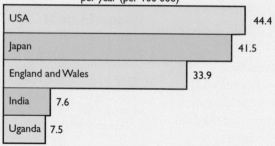

USA	44.4
Japan	41.5
England and Wales	33.9
India	7.6
Uganda	7.5

a) In which two countries is bowel cancer most common?

b) Draw a bar chart to show the data in the table below.

Country	% food high in fibre
England and Wales	64
India	87
Japan	85
Uganda	95
USA	64

c) In which two countries do people eat the most fibre-rich food? What is the incidence of bowel cancer in these countries?

d) In which two countries do people eat the least fibre-rich food? What is the incidence of bowel cancer in these countries?

e) Do your bar charts support the idea of a link between fibre in the diet and bowel cancer?

5 a) Copy the diagram of the human gut below. Use the labels below in place of the labels A–H on the diagram:

mouth; oesophagus; stomach; liver; pancreas; small intestine/ileum; large intestine/colon; anus

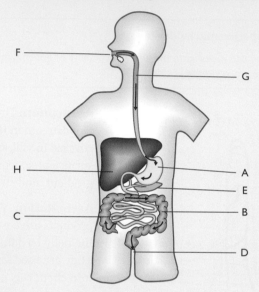

b) Annotate your diagram to give the function of each part of the gut that you have labelled.

6 Copy and complete this table to show the main digestive enzymes, where they come from in the gut, and what they do.

Name of the enzyme	What the enzyme works on	Products of digestion
protease		
carbohydrase		glucose
	fats	glycerol and

7 The liver produces bile, which is stored in the gall bladder and released into the small intestine when it is needed.

a) How does bile maintain the pH of the small intestine?

b) Bile emulsifies fats.

  i) What does this mean?

  ii) Why is it important?

8   This diagram shows villi in the small intestine. Digested food is absorbed by the villi and passed into the blood to be transported around the body. How does the structure of the villi enable food to be absorbed effectively in the small intestine?

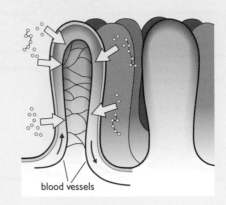

blood vessels

Chapter 5: Blood and circulation

- The body transport system consists of the blood vessels (the pipes), the heart (the pump) and the blood (the medium).

- Human beings have a double circulation – the pulmonary circulation to the lungs and the systemic circulation to the body.

- The three main types of blood vessels are the arteries, veins and capillaries.

- The heart is made mainly of muscle. It pumps blood around the body.

- Valves control the flow of heart in the blood.

- The blood has four main components – red blood cells, white blood cells, platelets and plasma.

- Blood clotting prevents blood loss and protects against the entry of pathogens.

- In vaccination the immune system is stimulated and memory cells are made, so that future exposure to a pathogen results in rapid production of large numbers of the right antibody to prevent disease.

Transport in humans

People have a very small surface area to volume ratio, so they need a transport system to carry sugar and oxygen to cells for cellular respiration. Waste products such as carbon dioxide and urea must also be removed. The heart pumps blood around a series of blood vessels.

Humans have a **double circulation**. The heart pumps blood to the lungs to pick up oxygen (**pulmonary circulation**). Oxygenated blood flows back to the heart. Then it is pumped out around the body (**systemic circulation**).

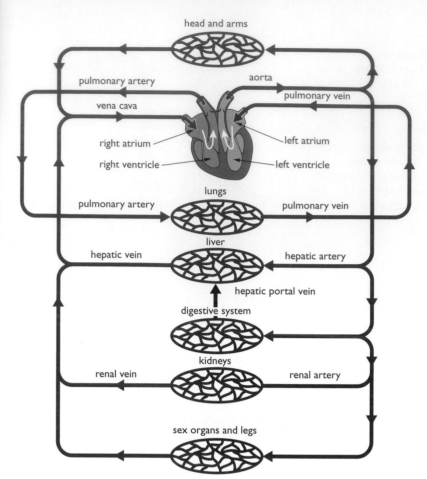

head and arms

pulmonary artery   aorta
                          pulmonary vein
vena cava

right atrium   left atrium
right ventricle   left ventricle

lungs
pulmonary artery   pulmonary vein

liver
hepatic vein   hepatic artery

hepatic portal vein

digestive system

kidneys
renal vein   renal artery

sex organs and legs

Figure 5.1 *The human double circulation*

space where blood flows   thick muscle layer

Arteries have thick muscular walls to smooth out the pulses of blood flow from the heart.

very thin wall

Capillaries have walls that are very thin so that the blood can exchange dissolved food and gases easily with the cells by diffusion.

large space for blood flow   thin muscle layer

Veins are large to carry blood under low pressure back to the heart.

Figure 5.2 *The structure of arteries, capillaries and veins*

The blood vessels

There are three main types of blood vessels – arteries, veins and capillaries. Arteries usually carry oxygenated blood away from the heart, except the pulmonary artery in the heart which carries deoxygenated blood to the lungs. Veins usually carry deoxygenated blood back to the heart from the tissues, except the pulmonary vein in the heart which carries oxygenated blood from the lungs to the left atrium.

Veins have valves. Substances diffuse in and out of the blood in the capillaries.

EXAMINER'S TIP ✔

Valves in the veins and heart stop the blood flowing in the wrong direction.

The heart

Blood enters the atria of the heart. These contract to force blood into the ventricles. When the ventricles contract, blood leaves the heart to go to the lungs (from the right) and around the body (from the left). Remember that both sides of the heart fill and empty together.

The heart beats about 70 times a minute. This beating is controlled by the **pacemaker region** of the heart. The rate of heartbeat changes as the demands of the body change. During exercise more food and oxygen are needed by the cells and more carbon dioxide is produced, so the heart beats more rapidly and more strongly.

Risk factors, such as a high fat diet, high blood cholesterol levels, obesity, etc., can lead to coronary heart disease. The coronary arteries which supply food and oxygen to the heart itself become narrower or blocked.

> **Experimental evidence**
>
> You can investigate the effect of exercise on the heart rate by measuring your own pulse rate under different conditions.

Blood pressure

When the chambers of the heart contract and force blood out into the arteries, the heart is in **systole**. When it relaxes and fills, it is in **diastole**. The blood pressure in the arteries changes as the heart empties and fills – it is higher as the heart pumps and empties, lower when it relaxes and fills. Normal systolic blood pressure is 120 mmHg. Normal diastolic blood pressure is 80 mmHg.

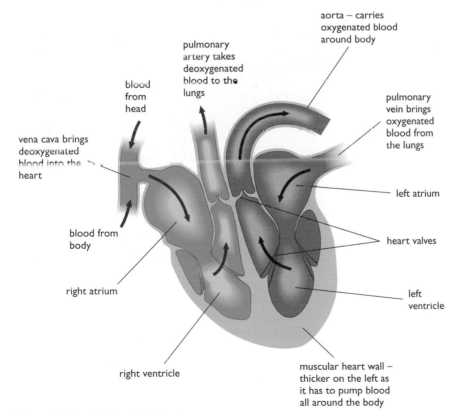

Figure 5.3 *The structure of the heart*

The components of the blood

- **Plasma** is the yellow liquid which transports dissolved food molecules, carbon dioxide and urea as well as all the blood cells. It is mainly water.

- Biconcave **red blood cells** contain haemoglobin and transport oxygen.

- **White blood cells** defend the body against attack by microbes. Lymphocytes produce antibodies to destroy microorganisms and memory lymphocytes give us immunity to specific diseases. Phagocytes engulf and digest microorganisms.

- **Platelets** are cell fragments which help clot the blood.

- Substances move by **diffusion** between the blood and the cells through the tissue fluid.

- Tissue fluid becomes lymph. It travels around the body in the **lymph system.** When it returns to the blood it is rich in antibodies against disease.

- **Blood clotting** involves several stages: injury → platelets arrive → platelets break open → if **calcium ions** present **thrombin** is formed → thrombin acts on **fibrinogen** to turn it to **fibrin** → fibrin forms mass of insoluble protein threads which forms a clot and then a scab as red blood cells become trapped in it.

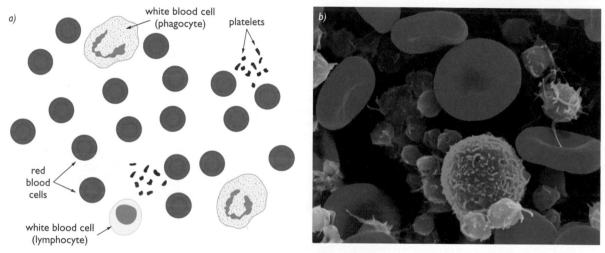

Figure 5.4 (a) *The different types of blood cells;* **(b)** *As seen in a photomicrograph*

The immune response

The white blood cells protect the body against pathogens. About 70% are phagocytes.

Pathogens have antigens on their cell surfaces. Lymphocytes make antibodies in response to these antigens. The antibodies stick to the antigens and destroy the pathogen in one of several ways:

- making the pathogens stick together so that phagocytes engulf them more easily

- acting as a label so that phagocytes recognise the pathogen more easily

- causing bacterial cells to burst open

- neutralising toxins produced by bacteria.

Some lymphocytes form memory cells so that if the pathogen gets into your body again it can be dealt with quickly before you are affected by the symptoms of illness. This secondary immune response is much faster and stronger than the first one.

You can be given **artificial immunity** to a disease by **vaccination.** You are given a weakened or dead strain of a pathogen. Your lymphocytes produce antibodies to the antigens of the pathogen, and memory cells ready for a secondary immune response, but you are not at risk of the disease. If you meet the live pathogen, your immune system will destroy it before you become ill.

Questions

1   Copy and complete these sentences. Use the words below to fill in the gaps.

circulatory	transported	heart	glucose
blood	waste products		oxygen

Substances are _____ round the body in the _____. Food molecules such as _____ and substances such as _____ are carried to the cells where they are needed. The blood also collects and carries away _____ from the cells. The _____ system is made up of the blood, the blood vessels and the _____.

2   This is a simple diagram of the double circulation of the human heart. Use it to help you answer the following questions:

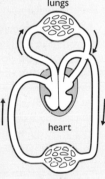

lungs

heart

rest of body

   a) Copy the diagram and shade it blue where the blood is deoxygenated and red where you would expect oxygenated blood.
   b) What happens to the blood in the body?
   c) What happens to the blood in the lungs?
   d) Why is it called a double circulation?

3   This diagram shows the main types of blood vessel.

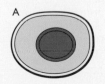
A

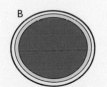
B

C

   a) Name the three types of blood vessel A, B and C.
   b) Describe the job of each type of blood vessel in the body.
   c) Explain how each vessel is adapted to its function.

4   Copy and complete each sentence, using the correct ending from below.
   a) Blood enters
   b) The atria contract and
   c) The ventricles contract and
   d) The blood leaving the right side of the heart
   e) The blood leaving the left side of the heart

Choose endings from:
   • force blood out of the heart.
   • is pumped to the lungs.
   • is pumped around the body.
   • force blood into the ventricles.
   • the heart through the atria.

5   Here are descriptions of two heart problems and how they may be overcome. In each case use what you know about the heart and the circulatory system to explain both the problems caused by the condition and how the treatment helps.
   a) Sometimes babies are born with a 'hole in the heart' – there is a gap in the central dividing wall of the heart. They may look blue in colour and have very little energy. Surgeons can close up the hole and the child can lead a normal life.
   b) The blood vessels supplying blood to the heart muscle itself may become clogged with fatty material. The person affected may get chest pain when they exercise or even have a heart attack. Doctors may be able to replace the clogged blood vessels with bits of healthy blood vessels taken from other parts of the patient's body, or open them up with a special metal tube called a stent.

6   Copy and complete the table on the next page to show the main parts of the blood, using the labels and annotations given below:

Part of the blood	Description
	The liquid part of the blood. Pale yellow and made mainly of water, it contains dissolved food molecules and other chemicals.
red blood cells	
white blood cells	
	These are small fragments of cells with no nucleus. They help to clot the blood.

- platelets

- These cells have a nucleus and help to defend the body against microbes which cause disease.

- plasma

- These cells have no nucleus. They are packed with the red pigment haemoglobin which carries oxygen.

7   The plasma is very important for transporting substances round the body. Three of the main substances are carbon dioxide, urea and digested food.

   a)   For each substance, say where in the body it enters the plasma.

   b)   For each substance, say where it is transported to, and what happens to it when it gets there.

8   The red blood cells carry oxygen around the body. They can do this because they are packed with a red pigment called haemoglobin, which picks up oxygen to form oxyhaemoglobin. This then carries the oxygen to places that need it and rapidly gives it up, forming haemoglobin and oxygen.

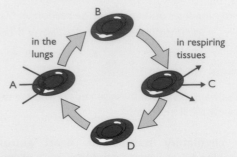

Label this diagram to show where in the body oxygen is picked up by the red blood cells and where it is unloaded.

9   The number of red blood cells in 1 mm^3 of normal human blood is about 5000 million. However, in certain situations the number of red blood cells in the blood may be particularly high or low. For each of the examples below, explain the difference in the red blood cell count and the effect it will have on the person.

   a)   People who live at high altitude (where the air contains less oxygen) have many more red blood cells than those living at low altitude.

   b)   An athlete who has used the illegal practice of 'blood doping' will have a higher than normal red blood cell count. Blood doping involves taking your own blood and storing it for several weeks, then having a blood transfusion of your own blood the day before a competition.

   c)   A person who does not have sufficient iron in their diet will be anaemic, feeling very tired and lacking in energy. Their red blood cell count will be lower than normal.

10 *a)*   Phagocytes and lymphocytes both help defend your body against disease. Explain the different ways in which they work.

   b)   How does vaccination use your immune system to protect you against a disease you have not had?

Chapter 6: Coordination

- Many multicellular organisms have both nervous and hormonal coordination and control systems.

- The central nervous system (CNS) consists of the brain and the spinal cord. They are linked to the sense organs and the effector organs by nerves.

- A nerve cell (neurone) consists of a cell body, dendrites and an axon.

- Sensory neurones carry information from the sense organs to the central nervous system (CNS).

- Motor neurones carry instructions from the CNS to the effector organs (muscles and glands).

- Neurones carry electrical impulses.

- The junctions between neurones are called synapses.
- A nerve contains many neurones. There are sensory nerves, motor nerves and mixed nerves.
- Reflex actions do not involve conscious thought.
- Reflex actions involve stimulus → receptor → sensory neurone → relay neurone in CNS → motor neurone → effector → response.
- Sense organs detect changes inside and outside your body.
- The function of the human eye is to receive and focus light to form an image. It contains many different parts, all adapted to their job.
- The pupil reflex controls the amount of light coming into your eye.

The human nervous system

Your central nervous system (CNS) – the brain and the spinal cord – gives you rapid responses to changes in your environment (**stimuli**). It uses the information from your sense organs to coordinate reflexes and other actions. Your **peripheral nervous system** runs all over your body. It is made up of the spinal nerves (from the spinal cord) and cranial nerves (from the brain). These nerves carry information from the sensory organs to your CNS, and also carry information from the CNS to the effector organs, which are muscles and glands. The sequence of events is:

stimulus → receptor → coordination → effector → response

An individual nerve cell is a **neurone**. A neurone responds to stimuli. It can conduct electrical impulses. A bundle of neurones is a **nerve**.

Synapses

Wherever one neurone ends and another begins, there is a tiny gap called a **synapse**. The electrical nerve impulse cannot cross this gap, so chemical transmitters are released. They diffuse across the gap. When they reach the next neurone, they bind to receptors on the membrane and a new impulse starts up.

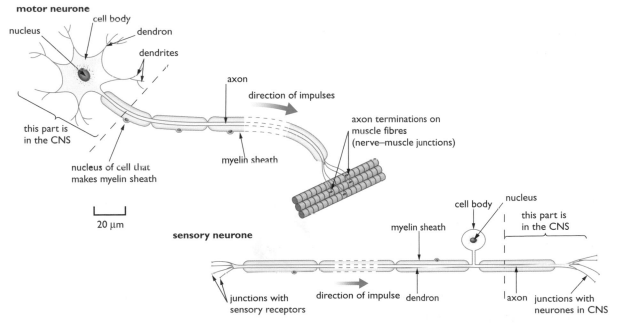

Figure 6.1 *Motor and sensory neurones*

Reflexes

Many of your actions are under conscious or voluntary control, e.g. you reach out for a piece of fruit. However, many are unconscious reflexes. Reflex actions are very fast. They protect you because they help you avoid danger, e.g. by pulling your hand away from something hot. They also run the everyday functions of your body, such as breathing. After a reflex has taken place, impulses travel to your conscious brain which make you aware of what has happened in the reflex reaction.

Spinal reflexes involve the spinal cord, for example when pulling your hand away from a hot object, or the knee-jerk reflex. Cranial reflexes involve the brain, e.g. the pupil reflex.

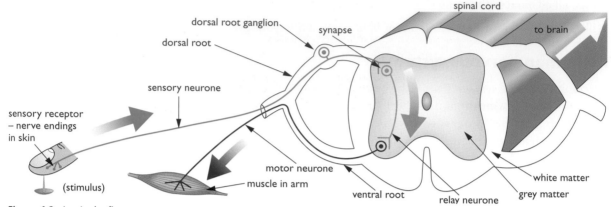

Figure 6.2 *A spinal reflex*

Sense organs

Sensory receptors are usually found in sense organs. They give you information about changes in the world around you and inside your own body. Sensory receptors transform energy from one form to another. The main sense organs are the eye (vision), ear (hearing and balance), tongue (taste), nose (smell), skin (touch, pressure, temperature and pain) and muscles (stretch receptors).

> **Experimental evidence**
>
> By using a bristle and a small grid, you can investigate the differing sensitivity of the skin on different areas of your arm and hand.

The eye is a good example of a sense organ.

> **EXAMINER'S TIP** ✓
> You need to be able to label a diagram of the eye and also use diagrams to show how the eye works.

Focusing

You focus your eyes by changing the shape of your lens. This is called **accommodation**. When you look at distant objects, the ciliary muscles relax and the suspensory ligaments pull tight. The lens goes thin as the light needs to be bent to focus on the retina. When you

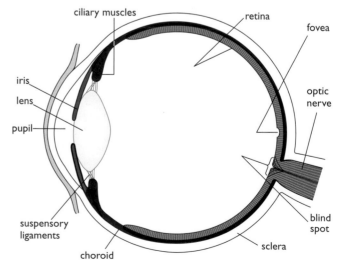

Figure 6.3 *The structure of the eye*

look at close objects, the ciliary muscles contract and the suspensory ligaments go slack. The lens is more spherical and fatter, as the light needs to bend much more to be focused on the retina. Revise the diagrams of this process.

The amount of light that enters your eye is controlled by the iris through the pupil reflex. The pupil enlarges in dim light to let as much light as possible into the eye. It gets smaller in bright light to protect the retina from damage.

The sequence of events in this reflex is stimulus (light intensity) → retina (receptor) → sensory neurones in optic nerve → unconscious part of brain → motor neurones in nerve to iris → iris muscles (effector) → response (change in size of pupil).

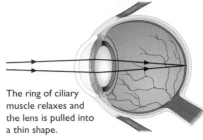

The ring of ciliary muscle relaxes and the lens is pulled into a thin shape.

Accommodation – how you see near objects

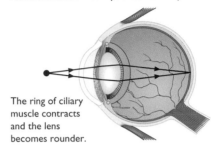

The ring of ciliary muscle contracts and the lens becomes rounder.

Figure 6.4 *Accommodation*

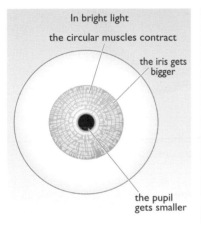

In bright light

the circular muscles contract

the iris gets bigger

the pupil gets smaller

So less light gets into the eye

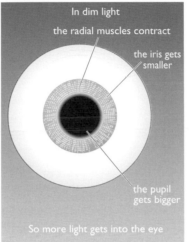

In dim light

the radial muscles contract

the iris gets smaller

the pupil gets bigger

So more light gets into the eye

Figure 6.5 *Pupil reflex*

The brain

The brain has many separate areas which control functions of the body or thought processes. The **cerebrum** controls our conscious thoughts, the **cerebellum** is involved in coordination and balance, and the **medulla** controls basic body functions. The **hypothalamus** and **pituitary gland** are parts of the brain involved in the chemical control of the body.

Questions

1 We have sense organs containing special receptors that allow us to detect changes in the world around us and inside our own bodies. Using this information, we can react to our surroundings. Copy and complete this table which shows some of our most important sense organs, using the words and expressions given on the next page.

Position of receptors	What are the receptor cells sensitive to?
eyes	
ears	
	changes in position – important for keeping balance
tongue	
	chemicals – enable us to smell
skin	

- light
- touch, pressure, pain and temperature changes
- ears
- sound
- nose
- chemicals – enable us to taste

2 Copy and complete these sentences. Use the words below to fill in the gaps.

controls	stimulus	coordinates
senses	nervous system	receptors

The human _____ enables your body to detect and respond quickly to stimuli. A _____ is a change in your surroundings or inside your body. Your _____ make you aware of these changes using special _____ which detect different kinds of stimuli. The rest of your nervous system _____ all the information and _____ the way your body responds.

3 The nervous system is made up of a number of parts. Explain the job of each of these parts:

a) the sense organs

b) the central nervous system

c) the sensory neurones

d) the motor neurones

4 a) What is the main difference between a voluntary action and a reflex action?

b) What is the value of reflex actions to the body?

c) Describe the following reflex actions using the sequence stimulus → receptor → sensory neurone → relay neurone in CNS → motor neurone → effector → response:

  i) a doctor hits you just below the knee cap with a rubber hammer

  ii) you put your bare foot down on a drawing pin

  iii) someone claps their hands near your face.

5 Look at the diagram below.

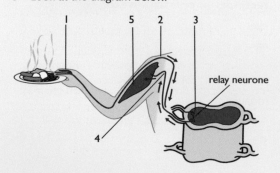

a) Write a description of what is happening at each of the numbered points 1–5.

b) How do you know consciously what has happened in a reflex action like this?

6 Parts of the eye have been labelled for you on the diagram below.

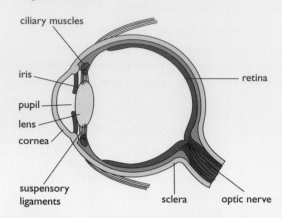

Copy and complete this table, matching the parts of the eye to the correct description of their functions.

Part of the eye	Function
	contains receptor cells that are sensitive to light
	tough white outer layer
	muscle that changes the size of the pupil
	curved, transparent area at the front of the eye
	focuses the image on the retina
	sensory neurones carry impulses from the retina to the brain
	hole that allows light into the eye
	change the shape of the lens
	attach the ciliary muscles to the lens

7 The pupil of the eye is the hole through which light enters. The size of the pupil is controlled by the muscles of the iris, and the pupil changes size depending on the light levels.

a) Draw diagrams to show what the pupil and iris of the eye would look like:

  i) in very bright light

  ii) in ordinary light levels

  iii) in very dim light.

b) Explain how the muscles of the iris change the size of the pupil.

8 Here is some information about three eye problems. Use what you know about how the eye works to explain why these conditions affect sight.

a) With cataracts, the lens goes cloudy or milky.

b) Some people have an eyeball that is more egg-shaped than round. They are often short-sighted – they can see close objects but not those at a distance.

c) If the retina of the eye becomes detached, the person goes completely blind in that eye.

9 Copy these diagrams and add to them, to help answer the questions below.

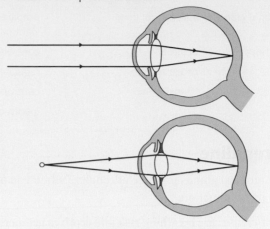

a) How do we focus on nearby objects?

b) How do we focus on distant objects?

Chapter 7: Chemical coordination

- Chemical coordination and control of the body is brought about by hormones.

- The hormones are secreted directly into the blood by special endocrine glands and carried around the body in the blood.

- Adrenalin and insulin are two examples of important hormones in the body.

- Adrenalin prepares your body for 'fight or flight'.

- Your blood glucose concentration is monitored and controlled by your pancreas.

- Insulin is a hormone that helps control your blood sugar concentration. Insulin converts glucose to glycogen.

- ADH (antidiuretic hormone), testosterone, progesterone and oestrogen are further examples of hormones.

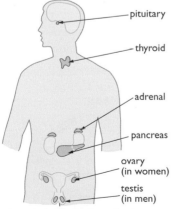

Figure 7.1 *The main endocrine glands of the body*

The endocrine system

The second system that coordinates and controls your body is the **endocrine system**. **Endocrine glands** make **hormones** which are released directly into the blood. These hormones are carried around the body in the bloodstream to the **target organs** that they affect. The target organs are often a long way from the endocrine glands that secreted the hormone. Hormones act as chemical coordinators.

The hormones secreted by the endocrine glands coordinate and control a wide range of bodily functions. Some of them are shown in the table.

Gland	Hormone	Functions of the hormone
Pituitary gland	antidiuretic hormone (ADH)	controls the water content of the blood by its effect on kidneys
Pancreas	insulin and glucagon	lower and raise blood glucose levels
Adrenal glands	adrenaline	prepares the body for stressful or physically active situations – 'fight or flight'
Ovaries	oestrogen	controls development of female secondary sexual characteristics
	progesterone	regulates menstrual cycle
Testes	testosterone	controls development of male secondary sexual characteristics
		involved in sperm production

Adrenaline

Your adrenal glands produce adrenaline when you are frightened, excited or angry. Adrenaline prepares your body for action.

- It increases the breathing rate and depth to get more oxygen into your body and remove excess carbon dioxide.

- It increases the heart rate and volume, sending more oxygen and food-rich blood to the muscles for respiration and removing more waste products.

- Blood is diverted from the digestive system to the muscles.

- Glycogen in the liver is converted to glucose in the blood.

- Mental awareness and the speed of reactions increase, and the pupils dilate, making your eyes more sensitive to movement.

- The body hair stands on end (useful to other species of mammals for making them look bigger).

Insulin and glucagon

Your blood glucose concentration needs to be around 4–6 mmol/l. This makes sure that your body cells always have the glucose they need for respiration to give them energy. It is kept at this level by insulin and glucagon (a hormone formed in the pancreas which helps break down glycogen, the carbohydrate stored in your liver).

EXAMINER'S TIP ✔

Make sure you're clear about the difference between **glucagon** (the hormone), **glycogen** (the carbohydrate stored in the liver) and **glucose** (the sugar carried around in the blood).

After a meal, blood sugar levels go up. Insulin is produced by your pancreas and lets your body cells take in some of this glucose. It lets your liver cells store glucose as glycogen. Then as blood sugar levels start to drop, insulin levels drop and glucagon levels rise. Glycogen is converted back to glucose to keep your blood sugar level up. The blood glucose levels are controlled by this negative feedback system involving the pancreas and the hormones insulin and glucagon.

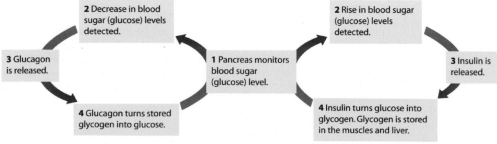

Figure 7.2 *The feedback system controlling your blood glucose levels*

Some people have diabetes, a disease in which the body does not make enough insulin to keep the blood glucose level constant. Blood glucose may rise to fatally high levels while the cells are starved of glucose for respiration. One symptom of diabetes is sugar being excreted in the urine. There is a simple chemical test for this. Diabetes can be treated by injections of insulin before meals. The amount of carbohydrate in the diet and the amount of exercise taken must also be carefully controlled.

Comparing nervous and hormonal control

The nervous system and the endocrine system both work to coordinate the functions of the body, but there are clear differences in how they do this.

Nervous control	Hormonal control
• Electrical messages travel along neurones • Chemical messages travel across synapses • Messages travel fast • Messages usually have a rapid effect • Usually a short-lived response • Nerve impulses affect individual cells, e.g. muscle cells, so have a very localised effect	• Chemical messages travel in the blood • Messages are transported slightly more slowly in the blood – minutes rather than milliseconds • Only chemical messages are involved • They often take longer to have an effect • Effects are often widespread in the body, affecting any organ or tissue with the correct receptors • Effects are often long lasting

Questions

1 Copy and complete these sentences. Use the words below to fill in the gaps.

blood	chemical	electrical	glands
long-term	nervous	rapid	

Your _____ system carries fast _____ impulses. It is an important part of the _____ response system of the body. Your hormones are _____ messages secreted by special _____ and carried around your body in the _____. Hormones are often part of the _____ control system of the body.

2 Match each word to its definition and then copy them out:

hormone   a hormone made in the ovaries that controls the development of the female secondary sexual characteristics

insulin   a chemical message carried in the blood that causes a change in the body

oestrogen   a hormone made in the pancreas that causes sugar to pass from the blood into cells where it is needed for energy

3 If you are excited, frightened or angry, your adrenal glands release adrenaline, the hormone that prepares you for 'fight or flight'. Adrenaline causes a number of changes to take place in the body.

a) List five of these changes.

b) For each of your answers, explain how it enables your body to deal with a threatening situation.

4

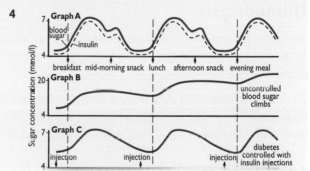

a) Look at graph A. Why does the level of insulin increase after a meal?

b) Graph B shows the blood sugar pattern of someone who has just developed diabetes and is not yet using injected insulin. What differences are there between this pattern and the one shown in A?

c) Graph C shows the effect of regular insulin injections on the blood sugar level of someone with diabetes. Why are the insulin injections so important to their health?

5 Some people become only mildly diabetic (type II diabetes). Their pancreas still makes insulin, but not enough to cope with the amount of carbohydrate and sugar-rich food they eat. This type of diabetes can be managed without needing to inject insulin. Explain how this might be done.

Chapter 8: Homeostasis and excretion

- You produce waste carbon dioxide (CO_2) when you respire. You produce waste urea when amino acids are broken down in your liver. You must balance the water and salt in your body so your cells do not have problems with osmosis.

- Water is lost through your lungs, your skin and your kidneys.

- Your kidneys filter your blood and control the amount of ions and water taken back into your blood (reabsorbed).

- The amount of water reabsorbed by your kidneys is controlled by a feedback system involving the hormone ADH from the pituitary gland in your brain.

- Your core body temperature is controlled by the thermoregulatory centre in the brain.

- If your core body temperature goes up, you lose heat through sweating and vasodilation. The body hairs flatten but in humans this has little effect.

- If your core body temperature falls, you need to make more heat. You also need to save heat to raise your body temperature. You stop sweating, vasoconstriction occurs, and your body hairs are raised to trap a layer of insulating air, though this does not help much in humans. Your metabolic rate goes up and you shiver to produce heat.

Homeostasis

You need to keep the conditions inside your body as constant as possible for you to survive. This is called **homeostasis**. Homeostasis is a coordinated response that needs a stimulus, a receptor and an effector to work.

Negative feedback is very important in keeping conditions constant inside your body. When levels in your body rise in response to a change in your environment, changes in your body lower the levels again. When levels fall, changes in your body make them rise again.

The kidneys

Excretion is one of the processes that make homeostasis possible. Urea is excreted through the kidney along with varying amounts of water and various solutes. It is these that are involved in the water balance of the body, an important part of homeostasis.

Every day you gain water in your food and drink and as your cells respire. You lose a similar amount of water, mainly in the urine but also through sweating, breathing out and faeces. It is important that your water input and output are balanced so that the internal environment of your body remains the same.

Your kidneys balance the amount of water left in your blood and filter substances out of your blood. They regulate the mineral ions in your blood, e.g. sodium ions, Na^+. You take in mineral ions with your food and lose some in sweat. You need some mineral ions in your blood. Your kidneys remove the excess ions from your blood and get rid of them through the urine. This balancing of water and salts is called **osmoregulation**. If the balance of water and mineral ions in your body is wrong, your cells can be damaged as water moves in or out of them by osmosis.

In the Bowman's capsule there is a process of **ultrafiltration**. The glomerular filtrate contains many things, including water, salt, urea, glucose and protein. The substances you need are taken back into your blood in the kidney tubules through a process called **selective reabsorption**. As the filtrate passes along the convoluted tubules and the loop of Henle, all of the glucose and protein is reabsorbed back into the blood.

The amount of mineral salts and water reabsorbed into the blood depends on the conditions of the body. This is controlled by **ADH (antidiuretic hormone)** from the pituitary, which affects the kidney in a **negative feedback system**. The liquid from the kidney tubules goes into the collecting ducts. ADH changes the permeability of the collecting ducts to water so that more or less water is reabsorbed into the blood. This changes the volume and concentration of the urine.

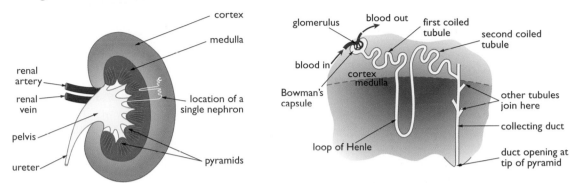

Figure 8.1 *The structure of the kidney and details of a kidney tubule*

ADH and osmoregulation

- Solute concentration of blood up → detected in hypothalamus → pituitary gland releases more ADH → ADH travels to kidney tubules in the blood → more water reabsorbed into blood by kidney tubules → small amount of concentrated urine formed, so less water lost → blood concentration of solutes restored to normal → ADH production falls to normal again.

- Solute concentration of blood down → detected in hypothalamus → pituitary gland releases less ADH → ADH travels to kidney tubules in the blood → less water reabsorbed into blood by kidney tubules → large amount of dilute urine formed, so more water lost → blood concentration of solutes restored to normal → ADH production rises to normal again.

Controlling body temperature

Your core body temperature needs to be around 37°C for your enzymes to work well. Your skin has many structures that can help control your body temperature. Sweat evaporates from the surface of your skin, transferring heat energy to your surroundings and cooling you down. When the blood vessels supplying your skin capillaries dilate, blood flows into the capillaries. You flush and lose heat by radiation through your skin. If those blood vessels constrict, the warm blood is kept deep in your body so that less heat is lost.

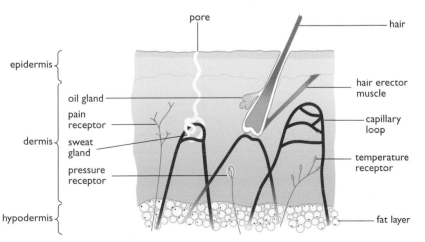

There is a negative feedback loop involving the **thermoregulatory centre** in your brain. It controls all the reactions of your skin to keep your body temperature stable.

Figure 8.2 *The structure of the skin*

Experimental evidence

Place a hand in iced water for several minutes to see some of the responses of the body to a drop in temperature. Then run hot water over your hand for several minutes to see some of the responses to a rise in temperature. Take care, the water must not be too hot or it might damage your skin.

Questions

1 Copy and complete these sentences. Use the words below to fill in the gaps.

kidneys	lungs	amino acids	urine
waste	respiration	urea	

The chemical reactions that take place in your body to keep you alive produce _____, which would be poisonous to your body if it were not removed. Carbon dioxide produced by _____ is removed through the _____ when we breathe out. When your liver breaks down excess _____ (from proteins), _____ is formed. This is removed from the blood by the _____ and excreted as _____ from the bladder.

2 If we take in a lot of liquid, the excess water is removed by our kidneys and we produce a lot of urine. Similarly, if we take in a heavy load of mineral ions, e.g. salt, our kidneys get rid of the excess. But our kidneys are not the only way we can control our water and salt loss.

a) In what two ways is water lost from our bodies other than in the urine?

b) How are excess salts lost from our bodies apart from in the urine?

c) How much glucose and protein would you expect to find in urine? Explain your answer.

3 The blood flowing into this kidney tubule contains, among other things, glucose, dissolved ions, urea and some traces of alcohol. Copy the diagram and make a table to explain what has happened to each of these substances at points A, B and C of the tubule.

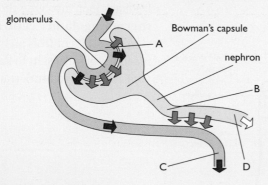

4 As part of an experiment, Jenny drank a given volume of water. Her urine output was then measured at 30-minute intervals for 150 minutes. The salt concentration of each urine sample was also measured. These graphs show what happened to both the volume of urine produced and the concentration of salt in the urine.

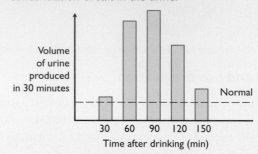

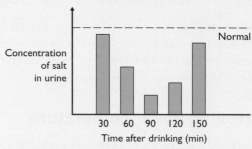

a) Describe what happened to the volume of urine produced by Jenny over the 150 minutes of the experiment.

b) Explain how the hormone ADH is involved in the changes in volume of the urine produced.

c) Describe what happened to the salt concentration of the urine over the same period.

d) Why do these changes in salt concentration happen?

e) What changes would you expect to see in the volume and salt concentration of the urine if Jenny's fluid intake were severely restricted and she did not drink for 12 hours?

5 Copy and complete these sentences. Use the words below to fill in the gaps.

constant	sweating	enzymes	die

It is very important to maintain a _____ body temperature of about 37°C. This is the temperature at which all the _____ in your body work at their best. If your body temperature gets too high or too low, you will _____. We cool down by _____ if our body temperature starts to go up.

6 Rearrange these sentences to show how Sally's internal body temperature is controlled when it starts to increase. Copy the sentences out in the right order.

  A Her body temperature starts to rise.

  B Sally takes a long, cool drink to replace the liquid she has lost through sweating.

  C Her temperature returns to normal.

  D Her skin goes red and sweating increases, so the amount of heat lost through her skin goes up.

  E Sally exercises hard.

7 Explain the role of

  a) the thermoregulatory centre in the brain and

  b) the temperature sensors in the skin

  in maintaining a constant core body temperature.

8 We maintain our body temperature at a constant level over a wide range of environmental temperatures. Many other animals – fish, amphibians and reptiles, as well as invertebrates – cannot do this. Their body temperature is always very close to the environmental temperature.

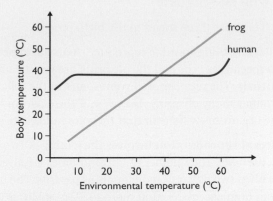

a) What is the body temperature of a person and a frog at an atmospheric temperature of 20°C?

b) From the graph, at what external temperature does the human core temperature become dangerously low? Why is it dangerous?

c) At what external temperature does the human core temperature become dangerously high? Why is it dangerous?

d) Explain how a person maintains a constant core body temperature as the external temperature falls.

e) Explain how a person maintains a constant core body temperature as the external temperature rises.

Chapter 9: Reproduction in humans

- Asexual reproduction involves only one individual. It gives identical offspring known as clones.

- Sexual reproduction involves two gametes from different parents. In fertilisation these sex cells fuse to form a zygote, a new, genetically unique individual. The zygote undergoes cell division and develops into an embryo.

- The sex organs in human beings mature and become active at puberty.

- The human male reproductive system produces sperm in the testes. They travel from the testes to the penis in the sperm duct, with secretions from the seminal vesicle and the prostate gland making semen.

- In the human female reproductive system, the ovaries release mature ova from developing follicles once a month in the menstrual cycle. The uterus develops a blood-rich lining each month to prepare for a pregnancy. If an ovum is fertilised, it will implant in the lining. If not, the lining is shed as the monthly period.

- The menstrual cycle in women is controlled by hormones released from the pituitary gland (FSH and LH) and by the ovary (oestrogen and progesterone).

- If the sperm and the ovum meet in the Fallopian tubes, one sperm may penetrate the ovum. This is the moment of **fertilisation** or **conception**.

- It takes around 40 weeks (nine months) for the fertilised ovum to grow and mature into a fully developed fetus ready for birth.

- There are three stages to the birth process.

All living things need to reproduce. **Asexual reproduction** involves only one parent organism. There are no special sex cells and the offspring are genetically identical to their parents. Asexual reproduction involves **mitosis**. The number of chromosomes in the cells remains the same. Reproduction is guaranteed and can produce many offspring. However, if the environment changes, the lack of variety in the genetic make-up of the organisms will mean that they are unlikely to survive.

Sexual reproduction involves the joining of two special sex cells or **gametes**. These sex cells are produced by **meiosis**. The number of chromosomes is halved when the gametes are formed. Sexual reproduction often involves two different individuals as parents. The offspring formed contain a mixture of genetic information from both parents. Sexual reproduction produces **genetic variety**.

In plants the sex cells are **ovules** and **pollen**. In animals the sex cells are **ova** and **sperm**.

Puberty

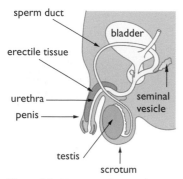

Figure 9.1 *Human male reproductive system (side view)*

Male secondary sexual characteristics such as the growth of the penis and testes, the growth of facial and body hair, muscle development and the voice breaking all take place during puberty. They are controlled by hormones. **FSH (follicle stimulating hormone)** and **LH (luteinising hormone)** are released by the pituitary gland. FSH stimulates sperm production while LH stimulates the testes to produce the male sex hormone **testosterone**. Testosterone causes the development of the male secondary sexual characteristics.

Female secondary sexual characteristics such as the development of breasts and body hair and the beginning of the menstrual cycle (periods) are also controlled by hormones. FSH (follicle stimulating hormone) and LH (luteinising hormone) are released by the pituitary gland. FSH stimulates the development of mature ova in the ovaries. They also interact with the female sex hormones oestrogen and progesterone.

EXAMINER'S TIP ✔
Remember that FSH and LH play a part in puberty in both boys and girls, but they affect different organs.

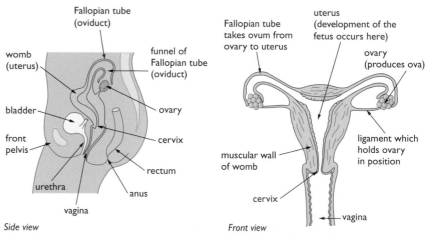

Figure 9.2 *Human female reproductive system (side and front view)*

The menstrual cycle

Following puberty, a woman will have a menstrual cycle approximately once a month. The menstrual cycles stop when she is pregnant and start again after the baby is born. They stop completely when she reaches the menopause. The menstrual cycle is controlled by hormones.

Pituitary hormones

- **Follicle stimulating hormone (FSH)** stimulates ova in the follicles of the ovary to develop. It stimulates the ovary to make **oestrogen**.

- **Luteinising hormone (LH)** stimulates the release of a mature egg from the ovary. It also stimulates the ovary to make **progesterone**.

Hormones from the ovaries

- **Oestrogen** stimulates the lining of the uterus to build up. It also reduces the levels of FSH produced by the pituitary and increases the levels of LH.

- **Progesterone** stimulates the growth of blood vessels in the lining of the uterus getting ready for a pregnancy. If a fertilised ovum arrives in the uterus, progesterone helps to keep the pregnancy going.

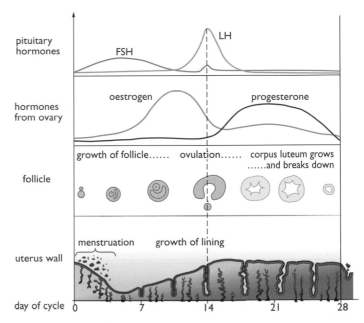

Figure 9.3 *The menstrual cycle*

Pregnancy

If a sperm fertilises an ovum and forms a zygote in the Fallopian tube, a pregnancy begins. The developing embryo moves into the uterus. It needs to be implanted in the lining of the uterus for the pregnancy to continue. As the embryo develops, the **placenta** forms. The placenta is a special organ that passes food and oxygen from the mother to the developing fetus. It also takes urea and carbon dioxide from the fetus and passes them into the mother's blood, and secretes the hormone progesterone which maintains the pregnancy. The fetus is connected to the placenta by the **umbilical cord**. Inside the uterus it is surrounded by amniotic fluid in the amniotic membranes. This cushions and protects the fetus as it develops.

The average length of pregnancy is 40 weeks. At the end of the pregnancy the baby is born.

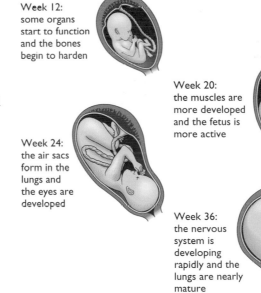

Week 12: some organs start to function and the bones begin to harden

Week 20: the muscles are more developed and the fetus is more active

Week 24: the air sacs form in the lungs and the eyes are developed

Week 36: the nervous system is developing rapidly and the lungs are nearly mature

Figure 9.4 *The fetus growing in the uterus*

1 Copy and complete these sentences. Use the words below to fill in the gaps.

> parent   asexual   gametes   different
> chromosomes   reproduction   sperm/ova
> sexual   identical   ova/sperm

There are two main types of _____.
_____ reproduction involves the joining of two special sex cells or _____. Each gamete contains half the number of _____ of the parent cells. In humans the sex cells are the _____ and the _____. The offspring are _____ from their parents. In _____ reproduction the offspring are genetically _____ to their one _____.

2 Puberty is the time when hormone-controlled changes take place in the bodies of boys and girls to produce sexual maturity.

a) What are the three main hormones that result in sexual maturity in boys?

b) What are the three main hormones that result in sexual maturity in girls?

c) Name three secondary sexual characteristics that develop during puberty in boys, and three in girls.

3 a)

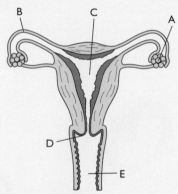

Copy the diagram of a front view of the female reproductive system. Use the labels below to replace A–E correctly.

> cervix   Fallopian tube (oviduct)
> ovary   uterus (womb)   vagina

b)

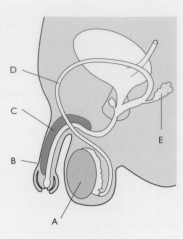

Copy the diagram of a side view of the male reproductive system. Use the labels below to replace A–E correctly.

> erectile tissue   penis   testis   seminal
> vesicle   sperm duct

c) Explain briefly but clearly:

i)   how the egg is fertilised by a sperm

ii) the role of the placenta during pregnancy

iii) how a baby is born.

4 If any of these parts of the reproductive system is damaged or does not work properly, a couple may be unable to have children. Explain clearly the role of each part and why infertility may result if it is damaged.

a) testis

b) ovary

c) oviduct (Fallopian tube)

d) sperm duct (vas deferens).

5 This is a diagram of a menstrual cycle.

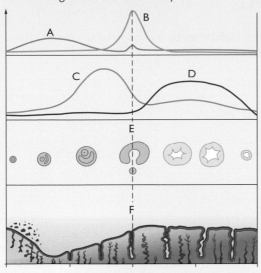

a) What is indicated by the letters A, B, C, D, E and F on this diagram of the menstrual cycle?

b) With reference to the different parts of this diagram, explain the events of the female menstrual cycle.

c) What differences would you expect to see if the woman became pregnant during this cycle?

6 Use your knowledge of the menstrual cycle and fertilisation to explain the following.

a) Fertile women can become pregnant only during a few days each month.

b) Young girls and old women do not become pregnant.

7 *a)* What are the functions of the placenta and how is it adapted to carry out these functions?

b) How would you expect the nutritional needs of a pregnant woman to change?

Chapter 10: Plants and food

- Photosynthesis is the process by which plants make food. Light energy is converted to chemical energy in the process.

- The word equation for photosynthesis is

$$\text{carbon dioxide} + \text{water} \xrightarrow[\text{chlorophyll}]{\text{light}} \text{glucose} + \text{oxygen}$$

- The chemical equation for photosynthesis is

$$6CO_2 + 6H_2O \rightarrow C_6H_{12}O_6 + 6O_2$$

- The rate of photosynthesis is affected by carbon dioxide concentration, light intensity and temperature.

- The structure of a leaf is adapted for photosynthesis and gas exchange with chloroplasts, a large surface area for diffusion of gases and stomata to allow the exchange of gases between the leaf and the air.

- Plants use oxygen and produce carbon dioxide during respiration in the cells. They use carbon dioxide and produce oxygen as a waste product during photosynthesis.

- Plants require mineral ions for growth, e.g. nitrate is needed to make amino acids to build proteins, and magnesium is needed to make chlorophyll.

Photosynthesis

The leaves of plants make glucose from carbon dioxide and water using light energy from the Sun. This energy is captured by chlorophyll in the chloroplasts of the leaves and converted to chemical energy in the glucose. Oxygen is also produced as a waste product.

Glucose is converted into sucrose to be transported around the plant, and into starch to be stored. Various experiments can be used to demonstrate what is happening during photosynthesis. You always start these experiments with a **destarched plant**, one that has been kept in the dark for at least 24 hours. Then you know that any starch produced is the result of photosynthesis during your investigation.

Experimental evidence

To show that photosynthesis has taken place demonstrate that starch is present in the leaf. Kill the leaf in boiling water, then boil in ethanol, rinse in water, and test the leaf for starch using iodine solution.

To show that photosynthesis is taking place by the presence of oxygen, use an aquatic plant such as Canadian pondweed (*Elodea*) in bright light. Capture the gas produced and analyse it to show raised oxygen levels, or simply count the number of gas bubbles given off.

To show that chlorophyll is needed for photosynthesis, use a variegated (green and white) leaf. This is of limited use because the products of photosynthesis could have been moved from one part of the leaf to another.

To show that carbon dioxide is needed for photosynthesis, remove carbon dioxide from the air surrounding one plant using soda lime. This is again of limited use because plants make carbon dioxide during respiration.

To show that light is needed, cover part of a leaf so that it is not exposed to light and show that starch is not made in that region of the leaf.

The glucose made in photosynthesis is used to make sucrose for transport, starch for storage, cellulose for cell walls, proteins and DNA, lipids as an energy store in seeds, and chlorophyll.

Adaptations of the leaf

Plant leaves have many adaptations to make photosynthesis as efficient as possible. They are thin and flat so as much light can be absorbed as possible and the distances that gases need to diffuse are very short.

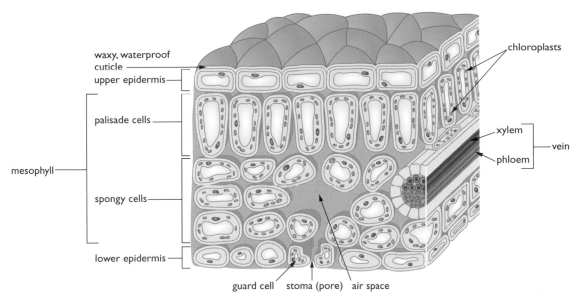

waxy, waterproof cuticle
upper epidermis
palisade cells
mesophyll
spongy cells
lower epidermis
guard cell   stoma (pore)   air space
chloroplasts
xylem
phloem
vein

Figure 10.1 *The structure of a leaf*

Photosynthesis and respiration

Plants respire all the time, using oxygen and producing carbon dioxide. The rate of photosynthesis varies depending on light levels, so the amount of carbon dioxide used up and oxygen produced varies during the day. Levels of carbon dioxide in the air around a plant vary during the 24 hours as the balance between respiration and photosynthesis changes.

Limiting factors

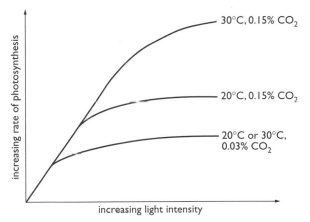

30°C, 0.15% CO_2

20°C, 0.15% CO_2

20°C or 30°C, 0.03% CO_2

increasing rate of photosynthesis

increasing light intensity

Figure 10.2 *Graph to show the effect of limiting factors on the rate of photosynthesis*

A limiting factor is the component of a reaction that is in short supply so that it limits the rate at which the reaction can take place. The main limiting factors for photosynthesis are the intensity of the light, the concentration of carbon dioxide and the temperature.

EXAMINER'S TIP ✔

Make sure you can explain limiting factors. Learn to interpret graphs that show the effect of limiting factors on the rate of photosynthesis.

Experimental evidence

You can investigate the effect of both light intensity and temperature on photosynthesis using pondweed.

Mineral nutrition in plants

Plants cannot survive on photosynthesis alone. They need mineral ions, which they take from the soil. This involves active transport as the mineral ions are taken up against a concentration gradient. Plants need nitrates to make amino acids, which are built up into proteins, and magnesium to make chlorophyll.

Questions

1 Copy and complete these sentences. Use the words below to fill in the gaps.

water	chloroplasts	starch	photosynthesis
light	chlorophyll	sugar (glucose)	

Plants produce their own food in a process known as _____. They absorb _____ energy using a green chemical called _____ which is found in _____ in the plant cells. Carbon dioxide and _____ are joined together using this energy to form _____ and oxygen. The glucose produced in photosynthesis may be converted into insoluble _____ for storage.

2 *a)* Match the words related to photosynthesis with the right description and then copy out the correct pairs.

Carbon dioxide gas   is produced and released into the air.

Water   provides energy.

Sunlight   from the root moves up to the leaf through the stem.

Sugars   is absorbed from the air.

Oxygen   are made in the leaf and provide the plant with food.

b) Write a word equation for the process of photosynthesis.

3 Copy this diagram. Use the labels provided to complete the labelling.

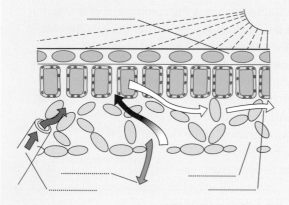

A Oxygen moves out

B The epidermis is a clear protective layer that lets sunlight in

C Sugars dissolve and are taken to other parts of the plant

D Stomata (pores) let gases in and out

E Water is brought from the roots in the xylem vessel

4 *a)* Much of the glucose made in photosynthesis is turned into an insoluble storage compound. What is this compound?

b) Some plants have their energy stores scattered about their stems and leaves. Other plants, like potatoes, develop special storage organs where they keep much of their stored food to help them survive the winter.

Amount of sunlight each month of the growing season	Mass of potato crop per plant (kg)
average	1
poor	0.5
high	1.5

One potato farmer recorded the amount of sunlight in each month that her potato crop was growing, and also recorded the average crop of potatoes she got from her plants when she harvested them. Explain her findings.

5 *a)* The glucose produced during photosynthesis is used for a number of things in the cells of the plant. List up to five different products made using the glucose produced during photosynthesis. For each product, explain how it is used in the plant cells.

b) The glucose made by photosynthesis is not only used to make other chemicals. What is the other main use of the glucose produced in the leaves?

6 Leaves are the plant organs that make food.

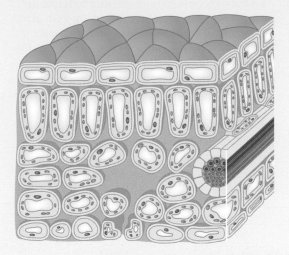

a) Copy the diagram of a section of a leaf and add the labels below:

A palisade layer

B spongy layer

C upper epidermis

D chloroplast

E stoma

F guard cell

b) Explain the function of each part you have labelled.

7 The diagram shows the apparatus required to demonstrate photosynthesis in *Elodea* (pondweed).

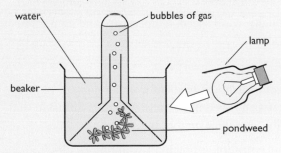

a) How do you know when photosynthesis is actually taking place?

b) You can use this basic apparatus to investigate the effect of light intensity on the rate of photosynthesis. What would you expect to happen if
i) you moved the light closer to the beaker containing the pondweed?
ii) you moved the light further away from the pondweed?

c) Why would you expect moving the light to have an effect on the rate of photosynthesis?

d) Which other factor might be changing each time you move the lamp?

8 The table shows the mean growth of two sets of oak seedlings that were grown in different amounts of sunlight.

Year	Mean height of seedlings grown in 85% full sunlight (cm)	Mean height of seedlings grown in 35% full sunlight (cm)
1966	12	10
1967	16	12.5
1968	18	14
1969	21	17
1970	28	20
1971	35	21
1972	36	23

a) Plot a graph to show the growth of both sets of oak seedlings.

b) Using what you know about photosynthesis and limiting factors, explain the difference in the growth of the two sets of seedlings.

9

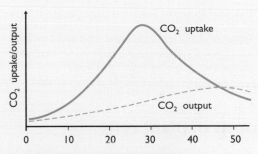

a) At what temperature is photosynthesis in the tomato plant at its maximum?

b) What effect does temperature have on the rate of photosynthesis in the tomato plant?

c) At what temperature is respiration in the tomato plant at a maximum?

d) To get the maximum yield of tomatoes, the plants need to be grown at their optimal temperature. This is the point at which the rate of photosynthesis exceeds the rate of respiration by the largest amount. Why will this result in the biggest crop of tomatoes?

e) At what temperature should a grower keep the greenhouse to get the maximum yield from tomato plants?

10 Plants need certain nutrients to grow well.

Nutrient	Part played in the plant
magnesium (Mg)	part of the chlorophyll molecule
nitrates (NO$_3$)	leaf and shoot production
	producing proteins – building blocks and enzymes
phosphates (P)	helps in the reactions of photosynthesis and respiration
	needed for healthy stems and roots
potassium (K)	good for flowers and fruit, and for disease resistance

At a plant clinic run by a fertiliser manufacturer, a number of farmers turn up with plants that are not growing as well as they should. For each plant, explain what is wrong with it and suggest what needs to be done to the soil to make sure that the crop picks up and grows well.

a) Plant A has yellow leaves with dead brown spots on them. Crops grown in the field the previous year produced poor flowers and little fruit.

b) Plant B shows stunted growth, and the older leaves have turned pale and yellow.

c) Plant C has poor growth of both stems and roots. The young leaves show a purple colour, while the older leaves are yellow.

Chapter 11: Transport in plants

- Osmosis is the movement of water down a concentration gradient of water through a partially permeable membrane.

- Water moves into and out of living cells by osmosis.

- Plant cells need to be turgid to support the plant.

- Water is taken up from the soil by the root hair cells and moves across the root by osmosis.

- Water and mineral ions are transported around the plant in the xylem.

- Sucrose and amino acids are transported from the leaves around the plant by the phloem.

- Transpiration is the evaporation of water from the surface of a plant.

- The rate of transpiration is affected by changes in humidity, wind speed, temperature and light intensity.

Osmosis in plant cells

Diffusion is the movement of solute molecules down a concentration gradient from a region of higher concentration of solute to a region of lower concentration. Osmosis is a specialised form of diffusion that takes place when two solutions are separated by a partially permeable membrane.

Osmosis is the movement of water molecules down a concentration gradient of water molecules from a dilute concentration of solute to a more concentrated solution of solute, through a partially permeable membrane. When only water can move through the membrane, it moves down a water concentration gradient.

Osmosis is very important in plant cells. The cell surface membrane is partially permeable. The cytoplasm and cell sap in the vacuole contain many solutes. If the plant cell is bathed in pure water or a dilute solution, water moves into the plant cell by osmosis. The vacuole and cell swell up and the cytoplasm pushes against the cell wall. The plant cell is **turgid**. This **turgor** is very important for plant cells to be able to support the structure of the plant.

If plant cells are placed in a solution more concentrated than the cell contents, water moves out of the cell by osmosis. The cell contents decrease in volume and the cytoplasm no longer pushes against the cell wall. The cell is **flaccid**. If the water loss by osmosis continues, the cytoplasm shrinks

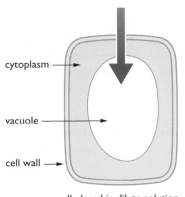

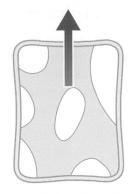
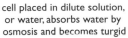

cytoplasm

vacuole

cell wall

cell placed in dilute solution, or water, absorbs water by osmosis and becomes turgid

cell placed in concentrated solution loses water by osmosis and becomes flaccid

excessive loss of water by osmosis causes the cell to become plasmolysed

Figure 11.1 *The effects of osmosis on plant cells*

right away from the cell wall. If plant cells lose water by osmosis, they cannot support the plant tissues and the plant **wilts**. Wilting helps protect the plant from further water loss.

Experimental evidence

You can investigate the effect of osmosis on plant cells using rhubarb or onion epidermis under the microscope or using cylinders or discs of potato tissue.

Plant transport tissue

Plants have two types of transport tissue (**vascular tissue**) which carry materials around the plant.

- **Xylem** tissue is made up of the cell walls of dead cells. It transports water and mineral ions up from the roots to the rest of the plant.

- **Phloem** tissue is made of living cells. It transports sugars for energy (particularly sucrose) and amino acids for cell building all around the plant.

In a young stem the vascular tissue is found around the edges. In the root it is in the centre.

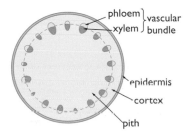

phloem
xylem
vascular bundle

epidermis

cortex

pith

Figure 11.2 *The arrangement of phloem and xylem in a young dicot stem*

Uptake of water and mineral ions by the roots

The outsides of the roots behind the growing tip are covered in root hair cells. These greatly increase the surface area of the root for the uptake of water. The concentration of mineral ions in the soil water is much lower than in the root cell cytoplasm. Water enters the root hair cells by osmosis. It moves across the root cells to the xylem along a concentration gradient by osmosis.

Mineral ions are at a very low concentration in the soil water. They are moved into the root hair cells against a concentration gradient. This is done using active transport, using energy from cell respiration. Once they are inside the root, the mineral ions move across the root mainly by diffusion and enter the xylem vessels to be carried around the plant.

EXAMINER'S TIP

Make sure you are clear about the differences between diffusion, osmosis and active transport when you are talking about moving substances in and out of cells.

Transpiration

Water leaves the leaf as water vapour through the stomata. It evaporates from the surface of the cells in the mesophyll, cooling the leaf. Water moves from the xylem to the mesophyll cells by osmosis. As water is lost by evaporation, the cytoplasm of that mesophyll cell becomes concentrated relative to the neighbouring cells, so water moves in by osmosis. This leaves those cells more concentrated and so water now moves into them from cells closer to the xylem.

The loss of water vapour from a leaf is called **transpiration**. It effectively 'pulls' water up the xylem all the way from the roots. This continuous flow of water is known as the **transpiration stream**.

The rate of transpiration

The rate of transpiration is controlled by the stomata. Most of the stomata are found on the underside of the leaf so as to reduce water loss. They open and close to allow gas exchange and to control the rate of water loss. Factors affecting the rate of transpiration, by controlling both the evaporation of water and the rate of photosynthesis, include light intensity, temperature, humidity and wind speed.

Experimental evidence

You can measure the rate of water uptake of a plant using a potometer. There are several different types. This gives an approximate measure of the rate of transpiration.

By changing the light intensity, air flow, temperature, etc., you can measure the effect of different factors on the rate of transpiration.

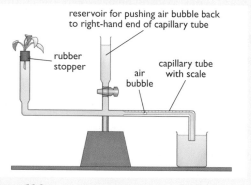

Figure 11.3

Questions

1 The diagram shows an experiment which is a model of osmosis in living cells.

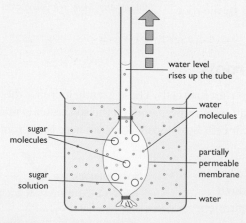

a) Explain what is happening in the experiment.

b) How is this model useful in explaining the entry of water into the xylem from the soil?

c) Using your knowledge of osmosis and diffusion, explain why plants store all their excess carbohydrate as insoluble starch.

d) Osmosis and diffusion take place along concentration gradients. When mineral ions such as nitrates are taken into plant roots, they are often moved against a concentration gradient. What is this called?

e) How does transpiration help in moving water up through the plant?

f) Why is it an advantage for the roots to have a large surface area?

2 Two students carried out an experiment on osmosis in plant cells using potato cylinders. They cut 30 equal-sized cylinders. The average length of the cylinders was 5 cm and the average diameter was 0.5 cm. Ten cylinders were put in distilled water, 10 in a weak sugar solution, and 10 in a strong sugar solution.

After several hours the students measured their cylinders again. Their results are given in the table.

Solution used	Average length of 10 cylinders at end of investigation (cm)	Average diameter of 10 cylinders at end of investigation (cm)
distilled water	5.5	0.6
weak sugar solution	5.0	0.5
strong sugar solution	4.5	0.4

a) Explain these results in terms of osmosis.

b) Draw what you would expect the potato cells to look like after several hours in each of the three solutions.

3 Copy and complete these sentences. Use the words below to fill in the gaps.

waxy   support   stomata   transpiration

Water inside plant cells provides the plant with _____. Plants lose water vapour by evaporation through the _____ (pores) in their leaves. This is known as _____. Most plants have a waterproof _____ layer on their leaves which stops them losing too much water.

4 Copy and label these diagrams, choosing the right labels to put in the boxes from those given in A–E:

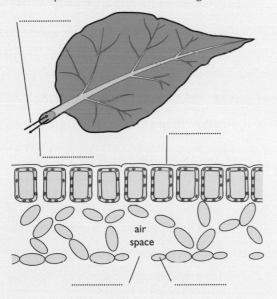

air space

A Xylem tissue in the veins brings water to the leaves

B Guard cells open and close the stomata, giving some control over water loss

C The waxy cuticle on the top surface of the leaf helps prevent unwanted water loss

D Phloem tissue in the veins takes sugars away from the leaf to all areas around the plant

E Stomata in the lower surface of the leaf allow carbon dioxide to diffuse into the leaf and water vapour to be lost by evaporation

5 Copy and complete these sentences, choosing the correct ending in each case.

Water loss by evaporation in a plant	in hot, dry and windy conditions.
Transpiration is more rapid	that the plant will lose too much water and wilt.
Plants keep relatively cool in hot sun	because transpiration cools them down.
Transpiration also creates a risk	is known as transpiration.

6 Plants make food in one organ and take up water from the soil in another organ. But both the food and the water are needed all over the plant.

a) Where do plants carry out photosynthesis?

b) Where do plants take in water?

c) There are two transport tissues in a plant. One is the phloem. What is the other?

d) Which transport tissue carries food about the plant?

e) Which transport tissue carries water about the plant?

f) What is the main difference between the cells of the two transport systems?

7 The diagram shows apparatus that can be used to give an idea of how much water a plant loses by transpiration.

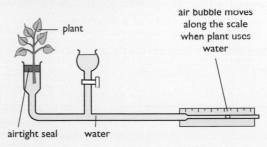

plant

air bubble moves along the scale when plant uses water

airtight seal   water

Use the diagram to help you answer the questions on the next page.

a) What is transpiration?

b) What part of the leaves helps to prevent them from losing too much water under normal conditions?

c) If the top surfaces of the leaves were coated with vaseline, how do you think it would affect the rate at which the plant takes up and loses water?

d) If the bottom surfaces of the leaves were coated in vaseline, how do you think it would affect the rate at which the plant takes up and loses water?

e) What do you think would happen to the air bubble in the capillary tube if you turned a fan onto the leaves of the plant? Explain your answer.

f) What is the apparatus in the diagram actually measuring?

Chapter 12: Chemical coordination in plants

- Plants respond to stimuli.
- Phototropisms are responses to the stimulus of light. Geotropisms are responses to the force of gravity.
- Roots and stems show geotropic responses.
- Stems have positive phototrophic responses – they grow towards the light.
- Roots usually grow away from the light.
- Stems have a negative geotropic response – they grow away from the direction of gravity.
- Roots have a positive geotropic response – they grow towards the direction of gravity.

Plant responses

Plants respond to the world around them. Their responses are usually much slower than animal responses, as they do not have a nervous system. Some plants can respond very quickly, such as the Venus fly trap.

Tropisms

Tropisms are **directional** responses by plants to stimuli from their environment. In tropisms, plants respond by growing towards or away from a stimulus.

- **Phototropisms** are the responses of plants to light coming from one direction.
- **Geotropisms** are the responses of plants to the force of gravity, which always acts downwards.

In *positive* tropism, growth occurs *towards* the direction of a stimulus. In *negative* tropism, growth occurs *away from* the direction of a stimulus.

Stimulus	Name of response	Response of shoots	Response of roots
light	phototropism	grow towards light source (positive phototropism)	most species show no response, but some grow away from light source (negative phototropism)
gravity	geotropism	grow away from direction of gravity (negative geotropism)	grow towards direction of gravity (positive geotropism)
water	hydrotropism	none	some species grow towards water (positive hydrotropism)

Phototropisms

Plants do not have sense organs in the same way as animals. Scientists have shown that the sensitivity of plant shoots to light is found in the growing shoot tip. The experimental work was based on oat coleoptiles – the first shoot that emerges from an oat seed as it germinates.

It can be shown experimentally that the stimulus for growth which brings about tropic responses to unidirectional light is a water-soluble chemical that is produced in the growing tip. It can be collected and used to make other shoots respond that have not been exposed to light from one direction. The chemicals involved are the plant hormones called **auxins**.

When unidirectional light hits a stem, auxins build up on the 'dark' side of the shoot. Auxins stimulate growth of the cells in the shoots. The more auxins, the more the cells grow. So the side of the plant with the most auxins grows more, which makes the shoot bend over towards the direction of the light.

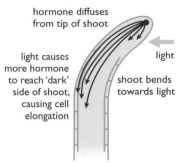

hormone diffuses from tip of shoot

light causes more hormone to reach 'dark' side of shoot, causing cell elongation

light

shoot bends towards light

Figure 12.1 *The response of a shoot to unilateral light*

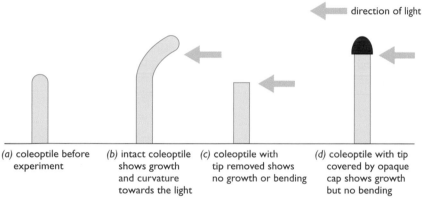

direction of light

(a) coleoptile before experiment

(b) intact coleoptile shows growth and curvature towards the light

(c) coleoptile with tip removed shows no growth or bending

(d) coleoptile with tip covered by opaque cap shows growth but no bending

Figure 12.2 *Simple experiments showing phototropisms in oat coleoptiles*

Experimental evidence

There are a number of experiments that can be used to investigate the response of plants to unilateral light. You need to grow some freshly germinated oat or wheat seeds to provide coleoptiles. You can remove or cover the shoot tips and apply artificial hormone to show how the responses come about.

Geotropisms

The way plants respond to gravity is not as well understood as the response to unidirectional light. Gravity always acts downwards on the plant. If a broad bean seedling is placed on its side in the dark, the tips of the shoots begin to grow upwards and the tips of the roots grow downwards. Auxin levels appear to be part of the response.

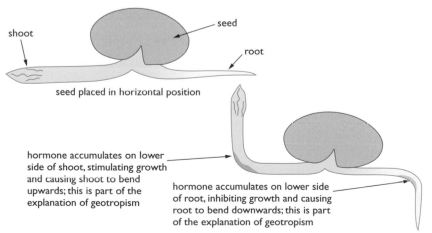

shoot

seed

root

seed placed in horizontal position

hormone accumulates on lower side of shoot, stimulating growth and causing shoot to bend upwards; this is part of the explanation of geotropism

hormone accumulates on lower side of root, inhibiting growth and causing root to bend downwards; this is part of the explanation of geotropism

Figure 12.3 *Using a clinostat to investigate geotropism in plants*

Experimental evidence

Geotropism can be demonstrated using a clinostat. This piece of apparatus holds the plant and slowly rotates in a vertical plane, so that gravity is sometimes acting downwards on any part of the plant, and sometimes acting upwards. This allows you to get rid of the directional stimulus of gravity so you can investigate geotropism.

1 Copy and complete these sentences. Use the words below to fill in the gaps.

| hormones | growing | stimuli | light | gravity |

Plants usually respond slowly to changes in their surroundings (called _____). They are sensitive to _____, moisture and the force of _____. Plants respond by _____ towards or away from a stimulus. The response is coordinated and controlled by _____ produced by the plant.

2 Plant shoots and roots respond differently to stimuli. Copy out these sentences, choosing the right word to describe how the plant responds.

a) Plant shoots grow **towards/away from** light.

b) Plant shoots grow **towards/away from** gravity.

c) Plant roots grow **towards/away from** gravity.

3

agar jelly

A   B   C   D   foil cap

glass strip

a) Copy these diagrams. Under each diagram, draw how you would expect it to look after two days.

b) Explain what has happened in each case, A–D.

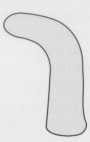

c) Copy this diagram of a shoot tip that has bent over in response to light from one side only. Add arrows and labels to the diagram to show the direction the light is coming from.

4 a) What stimuli do plants respond to?

b) What features of plant responses suggest that they are controlled by hormones rather than a nervous system?

5 Why are the responses of plant roots and shoots so important in the life of plants?

6 Explain clearly how you would demonstrate that plant shoots are negatively geotropic and plant roots are positively geotropic.

Chapter 13: Reproduction in plants

- Asexual reproduction involves only one parent. The offspring are identical to the parent. Sexual reproduction involves the joining of two special sex cells or gametes. The offspring are not the same as the parents.

- Plants can reproduce both asexually and sexually.

- Pollen needs to be carried from one flower to another. Flowers can be pollinated by the wind or by insects.

- After pollination a pollen tube grows down into the ovary to carry the male nucleus to the ovules.

- Fertilisation is the fusing of the male and female gametes. After this, seeds and fruits form.

- Seeds need warmth, water and oxygen to germinate.

- Seeds use their stored food as they germinate until the new leaves open and photosynthesise.

Asexual reproduction in plants

Plants can reproduce asexually to form identical offspring in many different ways, e.g. runners. Asexual reproduction is the result of **mitosis**.

People can encourage plants to reproduce asexually by taking cuttings.

Sexual reproduction in plants

Flowers are the sex organs of plants. They contain **pollen** grains, which are the male gametes, and **ovules**, which are the female gametes. Pollen is formed in the anthers of the stamens. Ovules are formed in the ovaries. The gametes are formed by **meiosis**. They contain half the number of chromosomes of the normal plant cells.

For the gametes to meet, pollination must take place. Pollen grains are transferred from the anthers of a flower to the stigma. Flowers which rely on insects to pollinate them have different features from flowers that are pollinated by wind.

Feature	Insect-pollinated flower	Wind-pollinated flower
position of stamens	inside petals, so insects come into contact with them	outside petals, so wind can blow pollen away
position of stigma	inside petals, so insects come into contact with them	outside petals to catch pollen blowing in the wind
type of stigma	sticky, so pollen grains attach	feathery and sticky to catch pollen blowing in the wind
size of petals	large to attract insects	small
colour of petals	bright to attract insects	usually green
nectaries	present to attract insects	absent
pollen grains	large and sticky to stick to insects	small, smooth and light to carry in the wind

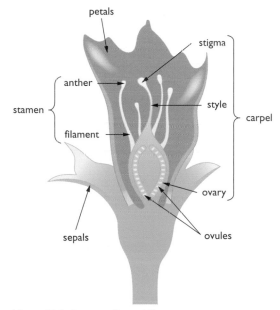

Figure 13.1 *Insect-pollinated flower*

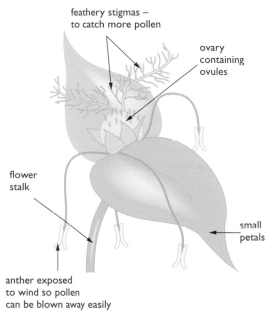

Figure 13.2 *Wind-pollinated flower*

Fertilisation

After pollination, the nucleus of the pollen grain must fuse with the nucleus of the ovule for fertilisation to take place. A pollen tube grows out of the pollen grain and down the style into the ovary into the ovule. The male nucleus travels down this tube from the pollen grain and fuses with the female egg nucleus in the ovule to form a zygote. This then develops to form a seed.

Seeds and fruit

Each seed contains the embryo plant, which has a root (the **radicle**) and a shoot (the **plumule**), the seed leaves or **cotyledons**, which may act as a food store for the young plant while the seed germinates, and the seed coat or **testa**.

The ovary wall becomes the fruit coat. There are many different types of fruit. These are adapted to **disperse** the seeds as far as possible from the parent plant. They may be dispersed by animals, wind or water. Dispersal is important because it avoids competition for resources such as water, minerals and light between the offspring, and between the parent plant and the offspring.

Germination

Germination is the process by which the embryo plant in the seed starts to grow to the point where it is photosynthesising independently. The radicle breaks through the testa and grows downwards into the soil (positive geotropism). The plumule grows upwards towards the light (negative geotropism).

The conditions needed for seeds to germinate are:

- warmth for maximum enzyme efficiency
- water to soften the testa and for chemical reactions in the seed to take place in solution
- oxygen for respiration to provide energy.

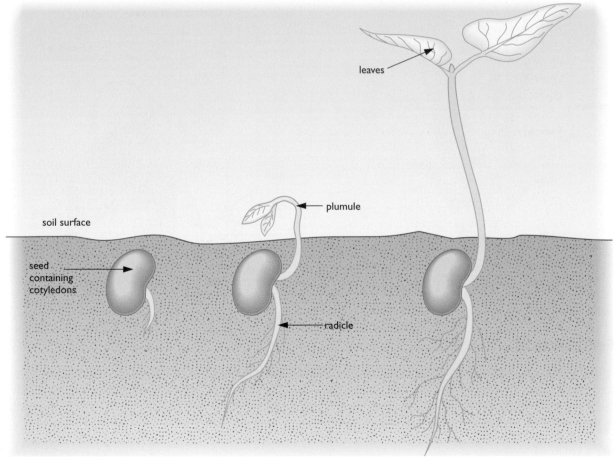

Figure 13.3 *The process of germination*

1 Copy and complete these sentences. Use the words below to fill in the gaps.

scent	asexually	pollinated	flowers
sexually	brightly coloured	wind	

Plants reproduce _____ and _____.
_____ contain the sex organs of plants. They can be _____ by the _____ or by insects. Insect-pollinated flowers are often _____ and produce _____ and nectar.

2 *a)* What are the main differences between asexual and sexual reproduction?

b)

A

i) Is structure A involved in asexual or sexual reproduction in the plant?

ii) What is this form of reproductive structure called?

iii) Explain how this structure is formed and its role in reproduction.

3 *a)* Asexual reproduction is fairly common in plants. Explain how it takes place in strawberries.

b) Asexual reproduction can be used artificially by gardeners to reproduce particularly good plants. One method of doing this is taking cuttings. Describe carefully the process of taking cuttings and explain why it is a form of asexual reproduction.

4 Copy and complete these sentences. Use the words below to fill in the gaps.

propagator	cuttings	light
transpiration	humid	

When gardeners take cuttings, they often put them in a special _____ or even in a polythene bag. This gives the cuttings a warm, _____ atmosphere. If they are also given plenty of _____, they have ideal conditions for successful growth. It is important for _____ to be in a humid atmosphere so they do not lose too much water by _____.

5 *a)* Copy and label this diagram of a flower.

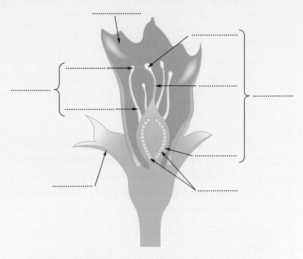

b) What is meant by the term pollination?

c) How do you think the flower in the diagram is pollinated? Explain your answer.

6 *a)* Draw and label a typical wind-pollinated flower.

b) What are the main differences between the structures of an insect-pollinated flower and a wind-pollinated flower? How are the structures related to their functions?

7 *a)* What is fertilisation?

b)

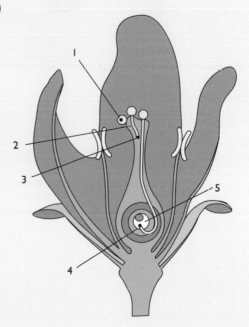

Look at the diagram of the fertilisation of a flower on the previous page. Match the labels on the diagram to the statements below, to describe the sequence of events that lead to the fertilisation of a flower in the correct order.

A The male nucleus fuses with the egg nucleus

B Pollen grains from another flower land on the stigma

C The male nucleus moves out of the pollen grain down the pollen tube

D A pollen tube grows out of the pollen grain down the style, and the pollen grain nucleus moves to the tip of the tube

E The pollen tube grows down and penetrates the ovule

8 Look at the fruits and seeds shown in the diagram below.

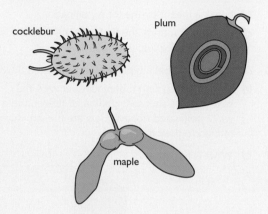

cocklebur

plum

maple

a) Why is seed dispersal so important to plants?

b) How do you think each of the fruits in the diagram is dispersed? Explain your answers.

9 Inside a seed, what are

a) the cotyledons?

b) the radicle?

c) the plumule?

d) the testa?

10 *a)* Describe three different ways in which plants can use the wind to disperse their seeds.

b) Draw a series of labelled diagrams to show what happens when a seed germinates.

11 *a)* What conditions do seeds need to germinate successfully, and why?

b) Choose one of these conditions and describe an investigation you could use to demonstrate that this condition is needed for successful germination.

12 Since we have found out about plant hormones we have developed a number of ways of using them to affect and control the ways plants behave. Explain carefully how we use plant hormones to produce large numbers of new plants quickly.

Chapter 14: Ecosystems

- An **ecosystem** is a distinct, self-supporting system of organisms interacting with each other and with their physical environment.

- A **population** is all of the organisms of a particular species living in an ecosystem at a particular time.

- A **community** is all of the populations of living organisms living in an ecosystem at a particular time.

- A **habitat** is the place where specific organisms live – their home.

- **Quadrats** can be used to sample the distribution of organisms in their habitats, and to estimate the population size of an organism in different areas.

- The feeding relationships of living organisms can be shown as **food chains**, **food webs**, **pyramids of number**, **pyramids of biomass**, and **pyramids of energy transfer**.

- Substances and energy are transferred along food chains.

- The different trophic levels of a food chain or web are known as **producers**, **primary**, **secondary** and **tertiary consumers**, and **decomposers**.

- Only about 10% of the available energy is transferred from one trophic level to the next.

- Carbon, nitrogen and water are all cycled through the environment and through living organisms.

Ecosystems

An **ecosystem** contains many living organisms that interact through feeding relationships. Plants are **producers** because they produce food by photosynthesis. Animals are **consumers** – they eat plants or other animals to get their energy. **Decomposers** are organisms such as bacteria and fungi that break down and decompose dead material and waste products to recycle the nutrients.

The population size of the organisms in an ecosystem is affected by the physical environment such as temperature, level of rainfall and amount of sunshine.

> **Experimental evidence**
>
> Quadrats are used to discover the numbers of plants or slow-moving animals in a habitat. You can use quadrats to estimate the population size of an organism in an area or to compare the distribution of organisms in different areas.

Organisms are continually interacting with each other in an ecosystem. Interactions include feeding relationships, competition between organisms, and relationships between the organisms and the physical environment.

Feeding relationships

The simplest way to show feeding relationships is with a food chain. Every food chain starts with a producer, for example

Grass → grasshopper → lizard

Plankton → crustacean → fish → seal → polar bear

Food chains oversimplify the situation. In a real ecosystem many food chains link together to form food webs. However, even food webs are an oversimplification. They do not tell you the numbers or mass of organisms involved, and they do not show the role of the decomposers.

Ecological pyramids

- **Pyramids of numbers** represent the numbers of organisms at each trophic level in a food chain. They have limitations because, for example, one tree can be home to many different plant-eating animals.

- **Pyramids of biomass** represent the total biomass of organisms at each trophic level in a food chain. Biomass is the mass of biological material that makes up an organism.

Here are two food chains and the pyramids of numbers and biomass for them. You can see the advantage of the pyramid of biomass – it always forms a pyramid and shows how the mass of biological material is less at each stage.

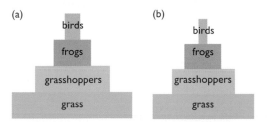

Figure 14.1 *Pyramids of numbers and biomass for food chain   grass → grasshoppers → frogs → birds*

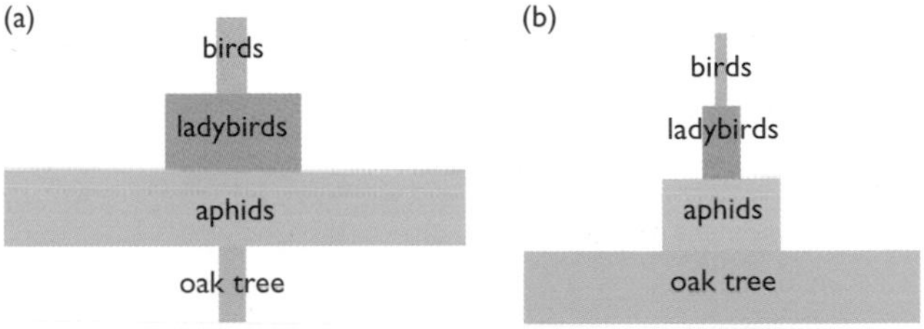

Figure 14.2 *Pyramids of numbers and biomass for food chain oak tree → aphids → ladybirds → birds*

The biomass of organisms gets less at each level, for the following reasons:

- Some parts of the organism are not eaten, e.g. plant roots, animal bones.
- The food cannot all be digested and absorbed – some is passed out as faeces.
- Some of the food eaten is broken down and excreted as waste, e.g. urea.
- Some of the food is respired and used to release energy for the cells of the body.
- Only a small part of the food is used to produce growth and new body mass.

Energy transfers between organisms

The interactions between the organisms in an ecosystem can also be shown as a transfer of energy along a food chain. Light energy is fixed into food molecules by plants during photosynthesis. Energy is released during respiration and used for movement, growth, reproduction, etc. Energy that is used for growth can be passed on to the next trophic level or the decomposers. Energy used for any other process will eventually be transferred to the environment as heat. Only about 10% of the energy is passed on at each level of a food chain. This limits the number of trophic levels in a food chain, because eventually there is not enough energy left.

Cycles in nature

In the **carbon cycle**, photosynthesis fixes carbon atoms from atmospheric carbon dioxide into organic molecules. These carbon-containing molecules are then passed along food chains by living organisms. Respiration by the living organisms releases carbon dioxide back into the atmosphere as organic compounds that are broken down to release energy. At the same time, fossil fuels formed from the remains of plants and animals store carbon compounds, and combustion of the fuels releases carbon dioxide back into the atmosphere.

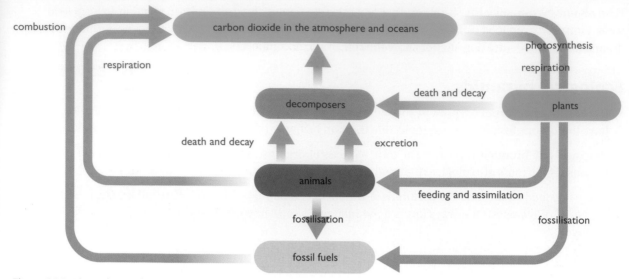

Figure 14.3 *The carbon cycle*

Nitrogen in the atmosphere cannot be used directly by most living organisms. The **nitrogen cycle** includes the following processes.

Nitrogen atoms in organic compounds are passed along food chains. Decomposition produces ammonia as proteins, amino acids, etc., are broken down. **Nitrifying** bacteria oxidise the ammonia to nitrites and then to nitrates in a process called **nitrification**. Plant roots can absorb the nitrates and use them to make organic molecules such as proteins.

Free-living nitrogen-fixing bacteria in the soil convert nitrogen gas into ammonia, which is then used to make proteins, etc. Ammonia is released into the soil when they die and decay.

Nitrogen-fixing bacteria in root nodules of leguminous plants make ammonia, which the plants use to make amino acids, etc.

Denitrifying bacteria use nitrates as an energy source and break them down into nitrogen gas.

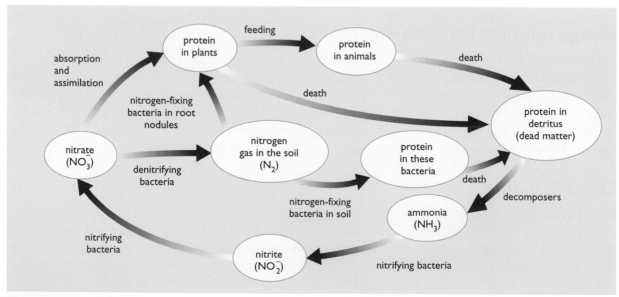

Figure 14.4 *The nitrogen cycle*

The **water cycle** is driven by heat from the Sun. Water is constantly recycled between the atmosphere, rivers, lakes and organisms and back to the sea. It involves evaporation from bodies of water, transpiration from plants, condensation in the atmosphere as the temperature drops, and precipitation (rain, snow, etc.).

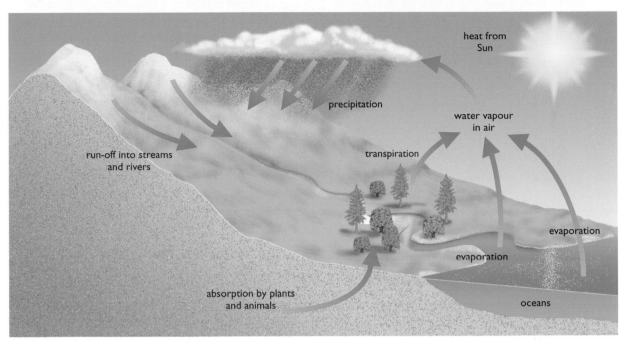

Figure 14.5 *The water cycle*

Questions

1 Match each word below with its correct definition, then copy out the correct pairs.

community	a group of organisms of the same species living in an area
habitat	all of the living organisms that share a habitat
ecosystem	the place in an ecosystem where an organism lives
population	a distinct self-supporting system of organisms interacting with each other and with their physical environment

2 Copy and complete these sentences. Use the words below to fill in the gaps:

consumers	**producers**	**secondary consumers**
animals	**plants**	**primary consumers**

_____ can make their own food using energy from the Sun, so they are known as _____. All of the _____ in the community rely either directly or indirectly on plants for their food – they are _____. Animals that eat plants are called _____ and animals that eat animals are _____.

3 Copy and complete these sentences. Use the words below to fill in the gaps:

animals	**producer**	**food chains**
photosynthesis	**energy**	

All living things need _____. Plants capture energy from the Sun and use it in _____ to make food. Animals get their energy by eating plants or other _____. _____ show us which organisms eat other organisms, and they always begin with a _____ (green plant).

4 Here are five jumbled food chains. Sort each one into the right order and then write them down.

a) stoat → primrose → rabbit

b) water fleas → stickleback (small fish) → tiny water plants → pike (big fish)

c) zebra → grass → lion

d) tiny sea plants → seal → fish → polar bear

e) blue tit → aphid → ladybird → rose bush

5 The following figure shows a number of interconnected food chains from the arctic tundra. Use it to help you answer the questions below:

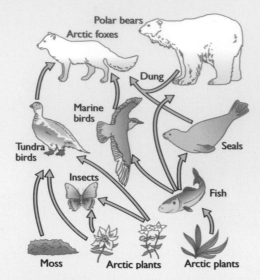

a) What do we call a number of interconnected food chains?

b) Which organisms are the producers in this system and why are they so important?

c) Which organisms are the primary consumers?

d) Which organisms are the secondary consumers?

e) Draw three food chains that make up part of this web.

f) What do you think would happen if a new disease appeared that killed most of the seals in the area?

6 a) Draw the pyramid of numbers for this food chain:
rosebush → aphids → ladybirds → birds

b)

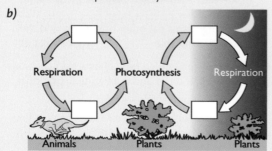

This drawing shows the same food chain as a pyramid of biomass. What do we mean by the term biomass?

c) What is a pyramid of biomass?

d) Using the example shown here, explain why a pyramid of biomass is often more useful than a simple pyramid of numbers.

e) Using the example shown here, explain why it is easier to use a pyramid of numbers than a pyramid of biomass.

7 At each stage of a food chain, less material and less energy are contained in the biomass of the organisms. Explain carefully:

a) how the energy/biomass is used by the organisms in a food chain

b) what happens to the energy that is *not* transferred from one level to another.

8

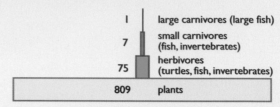

(values give g dry biomass per m²)

Figure 14.8

a) From this figure, calculate the percentage of biomass passed on

  i) from producers to primary consumers

  ii) from primary to secondary consumers

  iii) from secondary consumers to the top carnivores.

b) What is the average biomass transfer through this pyramid?

c) In any food chain or web, the biomass of the producers is much larger than that of any other level of the pyramid. Why is this?

d) In any food chain or web, there are only a small number of top carnivores. Use your calculations to help you explain why.

9 Draw a diagram of the carbon cycle showing plants, animals and carbon dioxide in the air. Add labels from the selection given below:

Photosynthesis: plants remove carbon dioxide from the air and store the carbon in the food they make

Respiration: animals give off carbon dioxide as they release the energy from their food

Respiration: plants give off carbon dioxide as they release the energy from their food

Decay: carbon dioxide is released by microbes, which decompose dead animals and plants and animal droppings

10 This diagram shows the nitrogen cycle.

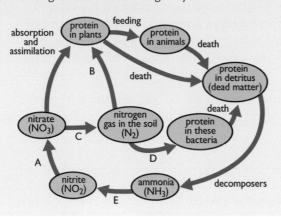

a) Why is nitrogen so important in living organisms?

b) Some plants, such as beans and clover, have special nodules on their roots that contain bacteria which can capture and 'fix' nitrogen from the air in a form that can be used by the plants. Why is this such an advantage for these plants?

c) Copy and label this diagram of the nitrogen cycle. Show clearly the role of microbes and detritus feeders in this cycle.

11 a) Why is water so important for life on Earth?

b) Explain carefully how the water cycle works.

Chapter 15: Human influences on the environment

- Human activities can affect the environment in many ways. Some of them are positive and some are negative.

- Glasshouses and polytunnels can be used to increase the yield of certain crops by controlling the levels of carbon dioxide and temperature.

- Fertiliser can be used on the land to increase crop yields.

- Pesticides can be used to reduce the damage done to crops by pests. There are advantages and disadvantages to using pesticides and biological control to deal with pests.

- Fish can be farmed to produce large numbers for food.

- The air can be polluted by **sulfur dioxide** which causes **acid rain**, and by **carbon monoxide**.

- **Greenhouse gases** trap heat energy from the Sun in the atmosphere around the Earth.

- The level of greenhouse gases in the atmosphere is increasing as a result of human activities and this is leading to **global warming**.

- Water can be polluted by sewage and leached minerals such as nitrates from fertiliser. Both can cause a lack of oxygen in the water. This is **eutrophication**.

- **Deforestation** has many damaging effects on the environment.

Food production

The human population of the world has grown enormously and is now over 6 billion people. This huge population puts many demands on the environment for food, building materials, fuel and space.

Food production affects the environment in many ways. Farmers are always trying to increase their yield (get more food from the same area of land).

- Farmers use greenhouses and polythene tunnels to grow crops. They can increase the rate of photosynthesis and so increase the crop yields. The glass or polythene gives a 'greenhouse' effect, warming up the growing area so that photosynthesis can take place as fast as possible to increase the yield. The level of carbon dioxide in the air can also be controlled to speed up the rate of photosynthesis.

- Fertilisers are widely used to increase crop yields. Fertilisers return nitrates and other minerals to the soil to make sure that plants have all the nutrients they need to grow as quickly as possible.

- Pests are controlled to prevent them damaging crops. Pesticides are the most widely used way of controlling pests. These are chemicals that farmers spray on their crops to kill the pests. They are very effective but they are expensive, pests can become resistant to them, and pesticides can cause environmental damage. They can also kill useful animals such as pollinating insects and the predators of the pests.

- Biological pest control uses another organism to control a pest. It can be very effective but it never completely gets rid of the pest. It reduces the numbers to a level where they are not causing significant crop damage.

Fish farming

More and more fish are being farmed to increase the amount of protein food available. Fish are kept in large enclosures so that the water quality can be monitored and controlled. The numbers of fish have to be carefully controlled to avoid too much competition between members of the same species. Other types of fish are kept out of the enclosures, to avoid interspecific competition. The quality of the food and the frequency of feeding are controlled, and selective breeding is used to produce fish that grow as fast as possible.

There are problems with fish farming. Disease can spread quickly and the waste from the fish and the excess food can cause water pollution, including eutrophication. Wild fish stocks are being depleted to make the pellets that are fed to farmed fish.

Air pollution

Pollution means releasing substances into the environment in quantities that cause harmful effects. We pollute the air with many gases.

Sulfur dioxide is produced when fossil fuels are burned. It combines with rainwater to form dilute sulfuric acid, which damages plants, kills life in streams and lakes, and damages buildings.

Carbon monoxide gas is formed when fuels are burned without enough oxygen. It combines quickly with the haemoglobin in the blood. Carbon monoxide can kill people if the blood becomes so saturated with the gas that the blood can no longer carry enough oxygen for cellular respiration to take place.

Carbon dioxide, water vapour, nitrous oxide, methane and CFCs are all greenhouse gases. They trap heat around the surface of the Earth and are needed to keep the planet warm enough for life. Human activities are leading to an increase in greenhouse gases. The greenhouse effect is enhanced and global warming results. This may lead to many environmental changes such as sea levels rising and changes in weather and rainfall patterns.

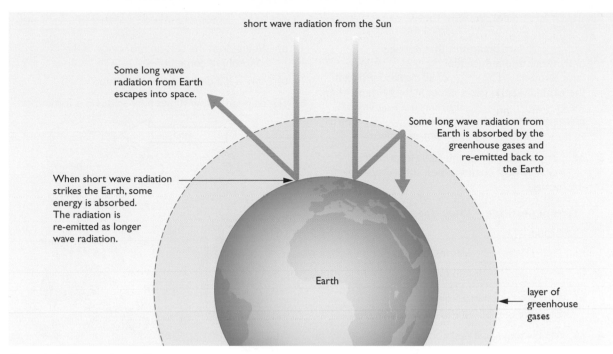

Figure 15.1 *The greenhouse effect*

Water pollution

The three main substances that pollute water are nitrates from fertilisers, sewage and detergents.

Nitrates **leach** from fertilisers in the soil into water, leading to eutrophication: nitrate levels up → increased algal growth → blocks light from water plants, which die → algae die as they run out of nitrates → decomposers break down dead plants and algae using oxygen → oxygen levels in the water fall and animals cannot live in it.

Sewage pollution also results in many more microorganisms that decompose the sewage. This uses up the oxygen in the water so that animals cannot survive in it.

Deforestation

Cutting down large areas of forest leaves the soil exposed. Minerals are leached out of the soil, the soil is eroded (washed or blown away), the water cycle is disturbed by lack of transpiration from the trees, and the balance of carbon dioxide and oxygen in the atmosphere is affected. Trees take in carbon dioxide and produce oxygen when they photosynthesise. After deforestation, carbon dioxide builds up in the atmosphere and less oxygen is made. In addition, the trees are often burned, using up oxygen and putting even more carbon dioxide into the atmosphere.

Questions

1 Farmers need to control the environment so that they can get the maximum yield from their crops and livestock. Explain clearly how the following farming methods help to increase yields:

  a) Application of inorganic (artificial) fertiliser to the fields

  b) Using a greenhouse or polythene tunnel for growing crops.

2 Copy and complete the sentences on the next page. Use the words below to fill in the gaps.

biological	stored	crops	pesticides
30%	pests	yield	

_____ are organisms that damage _____ or livestock and reduce the _____ from a farm. Around _____ of all the food grown in the world is lost to pests, either in the fields or when the crop is _____. Farmers use a variety of methods, from _____ to _____ control to try to prevent the yearly damage caused.

3 *a)* Copy and complete this table to show the main types of chemical pest control available to farmers.

Type of pesticide	What does it control?
herbicide	
	kills insect pests
	kills fungi

b) What are the main advantages and disadvantages of using chemical pesticides?

4 *a)* What is biological pest control?

b) What is the principle behind biological pest control?

c) Choose three methods of biological pest control and explain how they work: introducing a natural predator; introducing a herbivore; introducing a parasite; introducing a disease-causing microorganism; introducing sterile males; using pheromones.

5 One way of coping with the fall in fish stocks is by fish farming.

a) What species of fish are most commonly farmed?

b) How are fish farmed?

c) What are the disadvantages of farming fish?

6 Copy and complete these sentences. Use the words below to fill in the gaps.

damaging	local	pollution	population
organisms	human		

Human activities can have a big effect on the life of all other _____. The effect may be positive (for good) or it may be _____. When the _____ population of the Earth was much smaller, the effects of human activity were usually relatively small and _____. But the rapid growth of the human _____ and increased living standards mean that the effects of _____ can affect the whole planet.

7 Carbon monoxide is a colourless, odourless gas that is found polluting the air.

a) Where does the carbon monoxide in the atmosphere come from?

b) Why is it a dangerous air pollutant?

8 The diagram below shows how acid rain is formed.

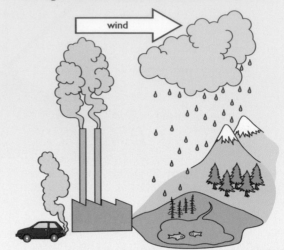

Copy and complete the diagram above. Use the labels below to help you:

A When fossil fuels are burned in cars or power stations, carbon dioxide is released along with sulfur dioxide and nitrogen dioxide.

B The gases may be blown in the wind to other areas or even other countries.

C The gases dissolve in the rain and make it strongly acidic.

D Acid rain damages trees directly, stripping the leaves and killing the plant.

E Acid rain damages water life indirectly – the water in lakes and rivers becomes acidic and this in turn kills plants and animals alike.

9 There are a number of gases in the atmosphere that play an important part in the greenhouse effect.

a) Name three important greenhouse gases.

b) Explain how the greenhouse effect works and why it is so important for life on Earth.

c) Increasing levels of some of these greenhouse gases have been measured over the last 50 years or so. Explain why the levels of greenhouse gases are increasing and how they are contributing to the greenhouse effect.

d) What is global warming and what effect might it have on life on Earth?

10 Read this information and then answer the questions below:

Water pollution is a serious problem in many places. One of the main pollutants is human sewage. As the human population grows, so does the amount of sewage produced. Much of the sewage, either raw or treated, is dumped into rivers and seas. Sewage is *biodegradable* – it can be broken down and used as food by microorganisms such as bacteria. Unfortunately, too much biodegradable material can cause a serious problem of *oxygen depletion* in the water.

Sewage is broken down into harmless chemicals in the water by the action of *aerobic* bacteria, which use oxygen dissolved in the water to break down their sewage 'food'. If there is plenty of sewage, the microorganisms increase rapidly. Unfortunately, oxygen is not very soluble in water – even water fully saturated with dissolved oxygen contains only about ⅕ of the concentration present in air. So if the numbers of aerobic bacteria get too high as a result of high levels of sewage pollution, they use up most of the dissolved oxygen. Other aquatic organisms that need oxygen, such as fish, then die of suffocation. The dead fish provide more food for the decomposing microorganisms, so the problem gets worse until the water can no longer support life.

a) What is meant by the term **biodegradable**?

b) Why is the pollution of water by sewage increasing?

c) Why does water pollution by sewage lead to an increase in the number of microorganisms in the water?

d) What is meant by the term **oxygen depletion** and why does sewage pollution lead to oxygen depletion of the water?

e) What effect does sewage pollution of the water supply have on other organisms such as fish and invertebrates in the water?

11 Cutting down large areas of forest is known as deforestation. Explain the effects of deforestation on the following:

a) soil erosion

b) mineral leaching

c) the water cycle

d) the balance of oxygen and carbon dioxide in the atmosphere.

Notes

Chapter 16: Chromosomes, genes and DNA

- The nucleus of the cell contains the chromosomes. The genes are located on the chromosomes.
- A gene is a section of a DNA molecule that controls the development of certain characteristics.
- A gene codes for a specific protein.
- Genes exist in alternative forms called alleles, which give rise to differences in the inherited characteristics.
- In human cells the diploid number of the cells is 46 and the haploid number is 23.
- The sex of a person is controlled by one pair of sex chromosomes, XX in females and XY in males.
- Mutation is a rare, random change in the genetic material that can be inherited.
- Many mutations are harmful, but many are also neutral and a few are beneficial.
- The incidence of mutations can be increased by exposure to ionising radiation or chemical mutagens.

Chromosomes and genes

Chromosomes are contained in the nucleus of the cell. They are made up of DNA, which is the molecule that carries all the inherited information. Small sections of the DNA control different inherited characteristics. These sections are known as genes and they control the production of specific proteins in a cell.

Each gene comes in at least two forms known as alleles. Each allele codes for a different form of the same character, e.g. free earlobes and attached earlobes.

Some alleles are **dominant** – the character they code for shows in the appearance of the individual, whether he or she inherits one or two of the alleles. Other alleles are **recessive** – the character they code for shows up in the appearance of the individual only if the allele is present on both chromosomes.

Many genes code for several different characteristics, but some code for single characteristics such as free or attached earlobes. These make it easier to study how characteristics are inherited.

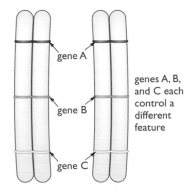

Figure 16.1 *An homologous pair of chromosomes*

Chromosome numbers

The normal body cells are **diploid**. They have 46 chromosomes each arranged as 23 pairs of matching (homologous) chromosomes. In 22 of the pairs both chromosomes are the same shape and size. The 23rd pair are the **sex chromosomes**. These decide the sex of the individual. Women have two similar X chromosomes (XX). Men have an X chromosome and a much smaller Y chromosome (XY).

The sex cells or gametes have only one of each pair. This means they have only 23 chromosomes each. They are **haploid**.

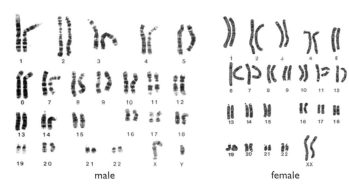

Figure 16.2 *A male and female karyotype showing the different chromosomes*

Mutations

A mutation is a change in the DNA of a cell. It can occur in an individual gene or involve a whole chromosome. Mutations take place when the DNA is replicating as a cell divides. Because the sequence of the DNA is changed, the gene might now code for a different sequence of amino acids and the protein might not be able to carry out its usual function.

Many mutations are harmful and affect the way a body organ works. Cancers are the result of mutations in cells as they grow and divide. If a mutation takes place in the sex cells, it may be passed on to the next generation. If it is a harmful mutation, this is usually known or recognised as a genetic disorder.

Many mutations make no difference, because the same protein is made. A few have positive benefits. Mutations occur randomly and naturally. Some factors increase the rate of mutation. These include ionising radiation (e.g. X-rays, ultraviolet light and gamma rays) and some chemicals (**mutagens**), including some of the components of cigarette smoke.

Whole chromosome mutations often result in the death of the cell. If chromosomes do not divide properly in a sex cell, too many or too few chromosomes might get passed on at fertilisation. If a fertilised egg has three copies of chromosome 21, for example, the new individual will have Down's syndrome.

Questions

1 Copy and complete these sentences. Use the words below to fill in the gaps.

> **characteristics** **offspring** **chromosomes**
> **parents** **genes**

Young animals and plants look like their _____. They have similar characteristics because of information passed on from parents to _____ in the sex cells from which they have developed. The information is carried by the _____ that make up the _____ found in the nucleus of every cell. Different genes control the development of different _____.

2 Match the words given below to the part they play in reproduction, then copy out the correct pairs.

nucleus — Its nucleus contains the chromosomes that carry genes from the father.

sperm — This contains the chromosomes carrying thousands of genes.

egg — Its nucleus contains chromosomes from both parents.

fertilised egg — Its nucleus contains the chromosomes that carry genes from the mother.

3 Link these words to their correct definitions, then copy out the correct pairs.

gene — an allele that controls the development of a characteristic even when it is present on only one of the chromosomes

alleles — an allele that controls the development of a characteristic only if it is present on both chromosomes

dominant allele — a unit of genetic information that is linked to a particular characteristic

recessive allele — different forms of a gene

4 The sex of a new human being is decided at the moment of conception.

a) How is sex determined in people?

b) The cells in about 50% of the human population show a dark patch inside, known as the Barr body. This appears because only one of their two X chromosomes is active in the cell; the other is deactivated and forms the Barr body.

Barr body

A group of young people were tested for the presence of a Barr body for experimental purposes. Twenty showed a Barr body and 24 did not.

i) How many of those tested were men?

ii) How can you tell?

5 Copy and complete these sentences. Use the words below to fill in the gaps.

genes	mutagens	variety
radiation	mutations	

New forms of genes arise from changes known as _____ in the existing _____. These occur naturally and are important for introducing _____. The chance of mutations occurring is increased by exposure to ionising _____ and chemicals known as _____.

6 Ionising radiation can come from radioactive substances such as uranium, ultraviolet light from the Sun, and X-rays.

a) When you are having an X-ray, the parts of your body that are not being irradiated are covered in lead, which absorbs radiation. Why?

b) People who work in X-ray departments move into a radiation-proof cubicle when they are using the X-ray machine. The levels of radiation they receive are carefully monitored. Why is so much care taken with people who work with ionising radiation?

7 Cigarette smoke contains a number of chemicals. Some of these are mutagens that make cells develop in an uncontrolled way.

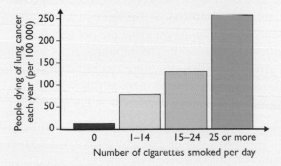

a) What effect does smoking cigarettes have on your chance of dying from lung cancer?

b) What effect does smoking cigarettes have on the levels of mutagenic chemicals in your body?

c) Why does smoking have a particular effect on the incidence of lung cancer?

d) In which other areas of the body would you also expect cancer to develop more frequently in smokers than in non-smokers? Explain your answers.

8 The table shows the effect of an increasing radiation dose on the number of mutations found in cells.

Radiation dose (arbitrary units)	Number of mutations per 100 cells
0.05	1
0.1	2
0.25	6
0.5	11
0.75	18
1.1	28
1.35	32

a) Plot a graph of these results, drawing a line of best fit through the points.

b) What does the graph show you about the effect of increasing doses of radiation on the rate at which mutations occur?

c) There have been concerns that people working with radiation in hospitals and nuclear power plants may be at increased risk of developing cancer. Why is this, and how can monitoring exposure to radiation reduce the risk?

9 Low doses of radiation may cause mutations to occur in the reproductive cells or in cells anywhere in the body. Explain carefully the long-term implications of:

a) mutations in the reproductive cells

b) mutations in normal body cells.

Chapter 17: Cell division

- The division of a diploid cell by mitosis produces two cells containing identical sets of chromosomes.

- Mitosis takes place during growth, repair, cloning and asexual reproduction.

- The division of a diploid cell by meiosis produces four cells, each with half the chromosome number of the original cell. This results in the formation of genetically different, haploid gametes.

- In humans the diploid chromosome number is 46 and the haploid chromosome number is 23.

- Random fertilisation produces genetic variation in the offspring.

- Variation within a species can be genetic, environmental or a combination of both.

All of the cells of your body apart from the sex cells are diploid – they have two sets of chromosomes. The sex cells are haploid, with only one set of chromosomes. The reason is that when the male and female sex cells join together in fertilisation, the new cell (or zygote) has the full number of chromosomes and can continue to grow and divide normally.

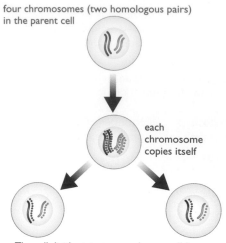

four chromosomes (two homologous pairs) in the parent cell

each chromosome copies itself

The cell divides into two; each new cell has a copy of each of the chromosomes.

Figure 17.1 *Mitosis*

Mitosis

There are two types of cell division, **mitosis** and **meiosis**. Mitosis results in two genetically identical, diploid, daughter cells. All of the cells in your body except for the sex cells are formed by mitosis from the zygote (the single cell formed when the egg and sperm meet at fertilisation). The sex cells are formed by meiosis.

In mitosis a copy of each chromosome is made before the cell divides. During the division each daughter cell receives a copy of each chromosome.

Mitosis is very important for the replacement of cells, e.g. skin cells, gut cells and blood cells. In cancer the normal control of mitosis is lost, so the cells divide very rapidly and repeatedly.

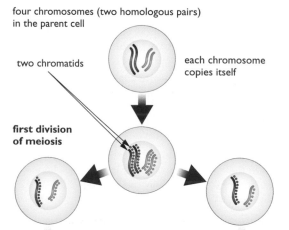

four chromosomes (two homologous pairs) in the parent cell

two chromatids

each chromosome copies itself

first division of meiosis

one chromosome (still containing two chromatids) from each homologous pair in each daughter cell

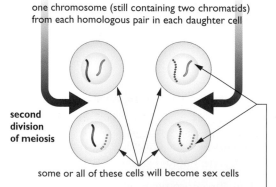

second division of meiosis

some or all of these cells will become sex cells

chromatids are separated and one chromatid from each chromosome ends up in each daughter cell

Figure 17.2 *Meiosis*

Meiosis

Meiosis is the special form of cell division that results in the sex cells or gametes. It involves two cell divisions and produces four haploid cells that are not genetically identical. In meiosis the chromosomes replicate to form two strands and then there are two cell divisions. The chromosomes are divided randomly between the daughter cells, so they are not all identical.

During the first division of meiosis, one chromosome from each homologous pair goes into each daughter cell. During the second division this chromosome separates into two strands, and one part goes into each daughter cell. This makes it possible to have a lot of variety.

In humans and other animals, meiosis is used to produce eggs and sperm. In plants it produces pollen grains and ovules.

EXAMINER'S TIP
Be very clear about the differences between mitosis and meiosis.

Sexual reproduction and variety

When two sex cells combine to form a new diploid cell (**zygote**) at fertilisation, genetic variety results in the offspring. This is because there is variety in the sex cells themselves as a result of meiosis, and then the fusing of the egg and sperm is also completely random. Every human being is genetically different from every other, except for identical twins who are formed from the same zygote.

Organisms show variety partly because of their genetic make-up. For example, pea plants can have alleles that make them grow either tall or short. But the height of the tall plants will vary because they will receive different amounts of light, carbon dioxide, water and minerals, so they will not all be able to grow to their full potential. Both genes and environment produce variation in organisms.

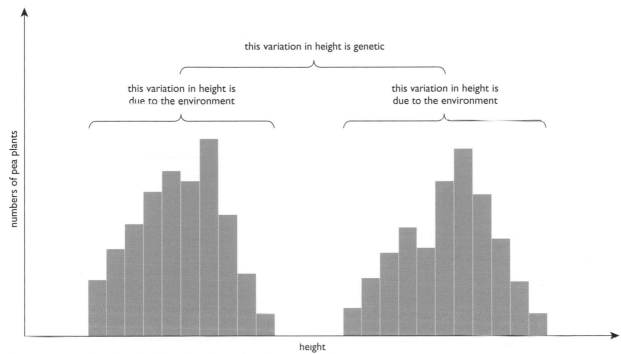

Figure 17.3 *Graph to show the effect of genetics and environment on variety*

Experimental evidence

You can investigate the effect of different conditions on growth by placing trays of germinating cress seeds in different conditions, or by taking several cuttings from the same geranium plant and keeping them in different conditions. In both cases compare the different specimens and record the variety and how it is caused.

Asexual reproduction and cloning

In asexual reproduction there are no sex cells and no meiosis. A part of the organism grows and forms a genetically identical new organism by mitosis. The identical offspring that result from asexual reproduction are known as **clones**. Asexual reproduction is very common in plants and some animals.

Section E: Variation and Selection

1 Sort out these jumbled sentences about sexual and asexual reproduction, and copy them out correctly.

In asexual reproduction no cells join	are known as gametes.
In sexual reproduction special male and female sex cells fuse (join)	contains a mixture of genetic information from both parents.
The new individual formed in sexual reproduction	to form a unique new cell.
The special sex cells involved in sexual reproduction	as a result of asexual reproduction.
A clone is the identical offspring formed	and the new individual is identical to the parent.

2

When the gametes fuse at fertilisation, they form a single cell. This cell divides to produce millions of cells with an exact copy of the chromosomes to form a baby. This cell division is called mitosis.

a) Copy and complete this diagram to show how mitosis happens. You will need to add chromosomes, labels and arrows.

b) Why is it so important that exact copies of the chromosomes are made?

c) Mitosis is very important during the development of a baby from a fertilised egg. It is also important all through life. Explain why.

3

a) How many chromosomes are there in a normal human body cell?

b) How many chromosomes are there in a human gamete (sex cell)?

c) What is the name of the special type of cell division that produces gametes from normal cells?

d) Whereabouts in the body would this type of cell division take place?

e) The figure shows this type of cell division. Copy and label it to explain what is happening at each stage.

4 Explain the roles of the different types of cell division in:

a) human reproduction

b) the production of new plants from cuttings.

5 Sometimes identical twins are separated at birth, adopted and brought up in different families. Comparing these genetically identical individuals who have been brought up in different environments with other identical twins who have been brought up together gives us some very useful information. We can also compare them with non-identical twins and with ordinary brothers and sisters (siblings).

Type of sibling	Difference in height (cm)	Difference in mass (kg)
identical twins brought up together	1.7	1.9
identical twins brought up apart	1.8	4.5
non-identical twins	4.4	4.6
non-twin siblings	4.5	4.7

a) Produce bar graphs to show these data clearly.

b) Which feature, height or mass, do you think is affected least by the environment in which identical twins are brought up?

c) Does the comparison made with non-identical twins and ordinary siblings confirm your answer to part **b**? Explain why.

d) This type of data on identical twins is very useful but quite rare. Why do you think such information is difficult to collect?

Chapter 18: Genes and inheritance

- Genes exist in alternative forms called **alleles** that give rise to differences in inherited characteristics.

- With a **dominant** allele, the characteristic will show up in the offspring even if only one of the alleles is inherited.

- With a **recessive** allele, the characteristic will show up in the offspring only if both alleles are inherited.

- If an organism is **homozygous** for a characteristic, both alleles are the same.

- If an organism is **heterozygous** for a characteristic, it has two different alleles.

- The **phenotype** describes the physical characteristics of an organism with respect to a particular pair of alleles (the genotype).

- The **genotype** describes the alleles that a cell or organism has for a particular feature.

- **Codominance** occurs when neither allele is dominant and both contribute to the appearance of the offspring.

- You can represent the patterns of single gene inheritance using a genetic diagram. You can use these to predict the outcome of various genetic crosses involving single genes.

- You can show the pattern of inheritance through a family using family pedigrees.

- The sex of a person is controlled by the sex chromosomes.

- You can use genetic diagrams to show how sex is determined when the egg and sperm meet at fertilisation.

How genetics works

The basis of genetics that we study today was worked out in the 19th century by a monk called Gregor Mendel. He carried out many careful genetic crosses using pea plants. An allele is inherited from each parent, but each parent has two possible alleles they could pass on. The combination of alleles that ends up in the offspring is the result of the random joining of the gametes at fertilisation. Genetic crosses can be represented by simple tables (*genetic diagrams*) that show you all the possible genotypes and phenotypes of the offspring.

EXAMINER'S TIP ✔
When you do genetic crosses, take care when choosing your letter to represent the dominant and recessive alleles of the gene. Use letters that are written differently in upper and lower case, e.g. A and a, G and g, rather than letters that look the same, e.g. C and c, W and w.

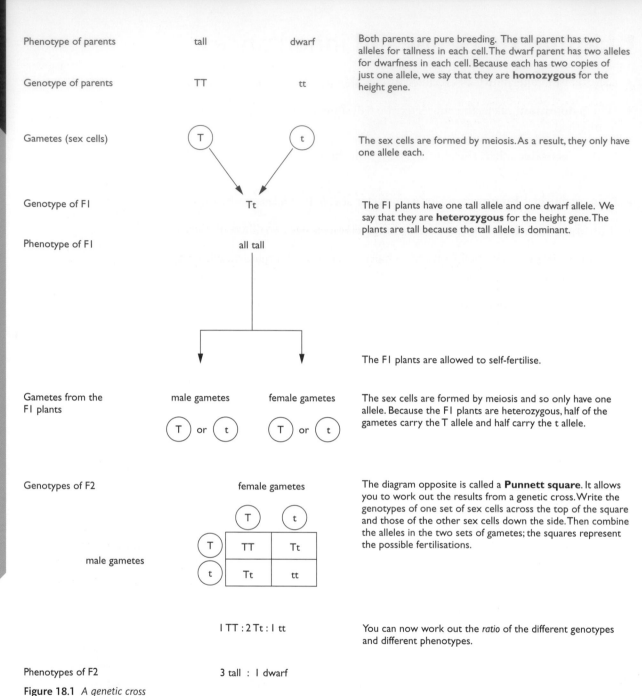

Phenotype of parents	tall	dwarf	Both parents are pure breeding. The tall parent has two alleles for tallness in each cell. The dwarf parent has two alleles for dwarfness in each cell. Because each has two copies of just one allele, we say that they are **homozygous** for the height gene.
Genotype of parents	TT	tt	

Gametes (sex cells)	T	t	The sex cells are formed by meiosis. As a result, they only have one allele each.

Genotype of F1	Tt	The F1 plants have one tall allele and one dwarf allele. We say that they are **heterozygous** for the height gene. The plants are tall because the tall allele is dominant.

Phenotype of F1	all tall	

The F1 plants are allowed to self-fertilise.

Gametes from the F1 plants	male gametes	female gametes	The sex cells are formed by meiosis and so only have one allele. Because the F1 plants are heterozygous, half of the gametes carry the T allele and half carry the t allele.
	T or t	T or t	

Genotypes of F2

The diagram opposite is called a **Punnett square**. It allows you to work out the results from a genetic cross. Write the genotypes of one set of sex cells across the top of the square and those of the other sex cells down the side. Then combine the alleles in the two sets of gametes; the squares represent the possible fertilisations.

female gametes

	T	t
T	TT	Tt
t	Tt	tt

male gametes

1 TT : 2 Tt : 1 tt

You can now work out the *ratio* of the different genotypes and different phenotypes.

Phenotypes of F2	3 tall : 1 dwarf

Figure 18.1 *A genetic cross*

Test crosses

It is impossible to tell just from its appearance (phenotype) whether an organism that has a dominant characteristic is homozygous or heterozygous for that trait. To find out, you carry out a test cross with an individual that is homozygous recessive for that characteristic. If the original organism is homozygous dominant, the offspring of the test cross will all show the dominant characteristic. If the original organism is heterozygous, half of the offspring will have the dominant characteristic and half the recessive trait.

Family pedigrees

Pedigrees are diagrams that show how a genetic characteristic or a genetic condition is passed on through generations of a family. Different symbols are used to represent males and females as well as affected and unaffected individuals. From a family pedigree it is often possible to work out the genotypes of the individuals involved, although sometimes you can only be sure of one of the alleles (see below). They are also known as family trees.

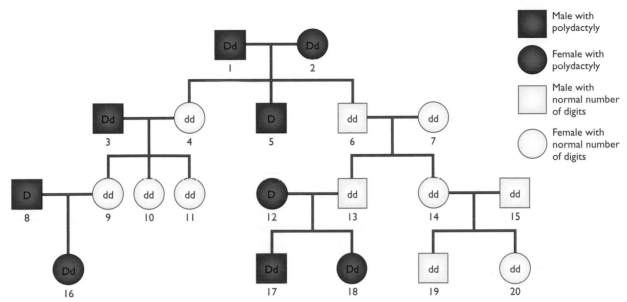

Figure 18.2 A family pedigree

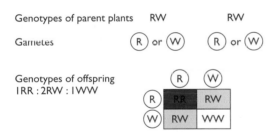

Genotypes of parent plants	RW	RW
Gametes	R or W	R or W

Genotypes of offspring
1RR : 2RW : 1WW

	R	W
R	RR	RW
W	RW	WW

Figure 18.3 An example of codominance

Codominance

In some cases, neither allele is dominant over the other and they both contribute to the final appearance of the offspring. An example is colour in snapdragon flowers – there is a red allele and a white allele. If a plant inherits one of each, it will be pink.

Sex determination

Phenotypes of parents	male	female
Genotypes of parents	XY	XX
Gametes	X and Y	all X

female gametes

		X
male gametes	X	XX
	Y	XY

Ratio of genotypes	50% XX : 50% XY
Ratio of phenotypes	50% female : 50% male

Figure 18.4 The inheritance of X and Y chromosomes

Sex is inherited on the sex chromosomes. Two X chromosomes make a person female, and X and Y make him male. Every egg contains an X chromosome. Half of the sperm contain an X chromosome and half contain a Y. The type of sperm that fuses with the egg determines whether the zygote and the individual will be female or male. Every time an egg is fertilised, there is an equal chance whether the zygote will be male or female.

1 Copy and complete these sentences. Use the words below to fill in the gaps.

recessive	genotype	homozygous
codominant	dominant	heterozygous
phenotype	characteristic	

A _____ individual has two identical alleles for a _____. If you have two different alleles, you are _____. Some alleles are _____ and show in the _____ even if only one allele is present. A _____ allele shows in the characteristics only if the _____ is homozygous. Some alleles both show in the phenotype – they are _____.

In Questions 2–8, it would help your answers if you include simple genetic diagrams to show the genetics of your explanations.

2 Whether you have dimples or not is decided by a single gene with two alleles. The dimple allele **D** is dominant to the non-dimple allele **d**. Use this information to help you answer these questions about Tom and Sandy. Tom has dimples but Sandy does not. They are expecting a baby.

 a) We know exactly what Sandy's dimple alleles are. What are they and how do you know?

 b) If the baby does not have dimples, what does this tell us about Tom's dimple alleles?

 c) If the baby has dimples, what does this tell us about Tom's dimple alleles?

3 Peas are usually round and green, but sometimes they are wrinkled. Peas have a gene controlling their appearance with two alleles, round and wrinkled. The round allele **R** is dominant to the wrinkled allele **r**. A gardener has been given a pea plant which has smooth, round peas. However, before breeding it with her pure-bred smooth, round peas, she wants to check that the plant is not heterozygous for the wrinkled gene by carrying out a test cross.

 a) What sort of plant would she need to carry out a test cross like this?

 b) Show the cross and the results she would get if the pea plant that she was given was

  i) homozygous

  ii) heterozygous for the smooth, round gene.

4 Manx cats are born with no tail. If a Manx cat is mated to a normal cat, on average half the litter will have tails and half will not. This suggests that Manx cats are heterozygous for a dominant allele **T** that causes no tail to form, with the normal tail allele **t** being recessive.

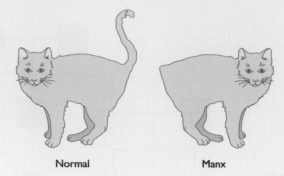

Normal   Manx

 a) Show the Manx cat cross described above.

 However, when two Manx cats are mated, there are always fewer Manx kittens born than we would expect. The homozygous form of the **T** allele is so damaging that the kittens that inherit it die before birth.

 b) Show a cross between two Manx cats. What ratio of Manx kittens to normal kittens would you expect to be born if the homozygous form was not so damaging?

 c) What ratio of Manx kittens to normal kittens is actually born, and suggest why this might be?

5 Cystic fibrosis is a genetic disease that particularly affects the lungs and gut. It is carried on a recessive allele.

 a) François and Annette are planning a family. François' sister has cystic fibrosis, and tests have shown that François is a carrier. Annette has had tests too, and she is not a carrier of the faulty allele. **F** represents the normal allele and **f** the cystic fibrosis allele. Produce a genetic diagram to show the chance of Annette and François having a child affected by cystic fibrosis.

 b) Steve and Paula also want to start a family. Neither of them has any relatives with cystic fibrosis. However, one or both of them could be a carrier of the cystic fibrosis allele. Produce genetic diagrams to show the different possibilities of their producing a child suffering from cystic fibrosis if neither has the affected allele, if one of them has the affected allele, or if both of them are carriers of the allele.

6

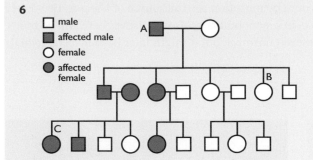

☐ male
■ affected male
○ female
● affected female

Achondroplastic dwarfism is a genetic condition that affects the long bones of the body, which do not grow to normal size, although in every other way affected individuals are quite normal. It is inherited as a dominant gene. Embryos that are homozygous for achondroplastic dwarfism die before birth. Use the family tree above to help you answer the following questions:

a) Choose a suitable capital and lower-case letter to represent the two alleles. Give the genotype you would expect for individuals A, B and C on the family tree.

b) In a family in which two people with achondroplastic dwarfism married, one pregnancy ended in miscarriage. What might be the explanation for this? Use a genetic diagram to help you explain.

7 You are given two plants of the same species, one with red flowers and one with white flowers.

a) What is codominance?

b) How would you use these two plants to investigate whether they show codominance in their flower colour? Produce genetic diagrams to show the possible results you might get.

8 A baby boy is born. He has dimples. Explain the difference between the way sex and dimples are inherited. Use genetic diagrams if you think they will make your explanations clearer.

Chapter 19: Natural selection and evolution

- Evolution takes place by means of natural selection.

- Natural selection can lead to change and adaptation within a population, which in time can lead to the formation of new species.

- Mutations – random changes in the genetic material that can be passed from one generation to the next – are rare.

- Some mutations are harmful, some are neutral, but a few are beneficial and give the organisms an advantage.

- Mutations can cause an increase in the resistance of bacterial populations to antibiotics.

What is evolution?

Evolution is the process by which the range of organisms on Earth change. New species continually arise from species that already exist, and some species become extinct – they die out completely. New species arise through a process known as **natural selection**. A species is a group of organisms that share common genes and can breed together to produce fertile offspring.

Living organisms always produce more offspring than are needed to replace them, but not all of these offspring grow up to become adults and breed themselves. Those organisms best suited to their local environment survive best and breed, passing on their genetic information to the next generation. This is natural selection. If conditions in an area change, the organisms that survive and breed will be those best suited to the new conditions. They may be different from the original organisms. In this way the different forms will become more and more different until eventually they are a new species.

Mutation

The random changes in the genetic material brought about by mutation are the source of the genetic variety that allows natural selection and evolution to take place. In any population of organisms, there will be many different mutations. Often a mutation will be neutral or even harmful, but it might later become beneficial if the environmental conditions change.

Natural selection in action

The dark form of the peppered moth is the result of a genetic mutation. It is at a big disadvantage in clean woodlands as it shows up against the light bark on the trees. However, when the UK environment changed after the Industrial Revolution and trees became blackened and polluted, moths of the dark form had an

Figure 19.1 *The dark and light forms of the peppered moth on different backgrounds*

advantage and their numbers in the population increased. This was an example of natural selection in action. As pollution has been reduced and the countryside has become cleaner again in recent years, the numbers of light moths have increased again.

The resistance of bacteria to antibiotics comes about as a result of random mutations. If bacteria are treated with the particular antibiotic that has been developed to fight a certain disease, any bacteria with a mutation that protects them against that antibiotic will survive and reproduce. If these bacteria then infect someone with the disease, they will not be destroyed by the normal antibiotic. This means the infection will no longer be cured by that antibiotic. An increase in resistance to antibiotics in bacterial populations makes it more and more difficult to control infections. The choice of antibiotics available which will destroy the bacteria causing the infection gets smaller. Organisms can become resistant to pesticides and other chemicals in the same way.

Natural selection and evolution

The examples of natural selection given above show you how organisms can adapt and change to new conditions as a result of natural selection. However, this process can continue indefinitely, eventually leading to the formation of completely new species because the organisms become so different they can no longer interbreed.

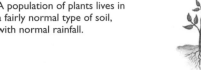

I A population of plants lives in a fairly normal type of soil, with normal rainfall.

2 Some of these plants colonise a different area, where water is found much deeper in the soil and the rainfall is considerably less. In this new environment, natural selection favours those plants with longer roots (able to reach the soil water) and smaller leaves with fewer stomata (to minimise water loss).

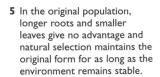

3 The two populations of plants are isolated from each other and cannot interbreed.

4 There is natural variation in these features in both populations as a result of sexual reproduction and gene mutation

6 In the new populations, longer roots and smaller leaves give an advantage, as plants with these features gain more water and lose less than those without them. In each generation, more plants with these features survive than those without them.

5 In the original population, longer roots and smaller leaves give no advantage and natural selection maintains the original form for as long as the environment remains stable.

7 The new plants become more and more different from the original population. Long-rooted and small-leaved forms survive best and the population eventually consists almost entirely of this type.

Eventually the two populations are so different that they cannot interbreed. At this point we consider them to be separate species.

Figure 19.2 *How natural selection can lead to the formation of different species*

Darwin and evolution

It was Charles Darwin who first developed the idea that all life on Earth has evolved from common ancestors by the process of natural selection. He developed his ideas based on the animals and plants he saw on his famous voyage of the *Beagle* along with lots of evidence he collected afterwards.

Questions

1 Random mutations may cause small changes in the way the body develops and works. The graphs show how the numbers of resistant insects in a population changed with the application of pesticide.

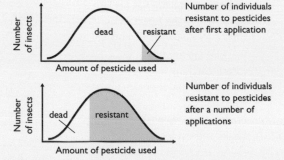

Explain how a pesticide can become ineffective against a type of insect as a result of mutations.

2 *a)* The table shows some of the evolutionary stages of an animal. Which animal is it?

	Height (ground to shoulder)	Body shape	Leg bones (digits numbered)	Habitat and leg shape
I	up to 1.6 m		3	Adapted to life in dry grasslands. Very efficient at running. Hoof formed from end of digit 3. Digit 3 lengthened.
II	1.0 m		3	Increased reliance on speed. Digits 2 and 4 very much reduced. Hoof formed from end of digit 3. Digit 3 thickened for support.
III	up to 1.0 m		4 — 2, 3	Very dry conditions: prairies. Speed more important. Digits 2 and 4 reduced. Digit 3 used for running – increased in length.
IV	up to 0.6 m		4 — 2, 3	Dry conditions: forests and prairies. Speed important to escape enemies. Only three digits very obvious. Digit 3 much enlarged.
V	about 0.4 m		5 — 2, 4 — 3	Size of fox. Lived on soft ground near streams. Four digits in forelimb and three in hindlimb for support on soft ground.

b) This is one of the best fossil records we have. Look carefully at the size of each animal and at the shape and arrangement of the foot. You also have some information about where we think each of these animals lived. Use this evidence to build up a theory of how these animals evolved through a process of natural selection, and why their feet changed in the way that they did.

3 Ptarmigans are birds, members of the grouse family, that are found in Scotland. Their summer plumage is a mixture of browns and cream, but after the autumn moult the feathers of most of the birds grow back almost white. A smaller number of ptarmigans keep their normal brown and cream colouring.

a) How is it possible for the winter plumage of different members of the same species of bird to be so different?

b) Is there likely to be a disadvantage in having brown feathers during a Scottish winter?

c) Do you think that having white feathers in the winter increases the breeding chances of those ptarmigans that change colour?

d) Does the fact that most ptarmigans have white plumage in the winter confirm your hypothesis?

4 One year, oyster fishermen in Malpeque Bay in Canada noticed that a few of the oysters they were catching were diseased and small, with pus-filled blisters. The graph shows the subsequent effect of this disease on the oyster population in that area.

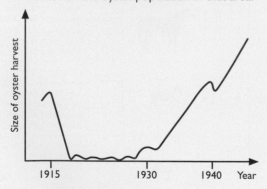

a) The first diseased oysters were noticed in 1915. How long did it take for the oyster harvest to be virtually wiped out?

b) In which year did the harvest really start to pick up again?

c) How long was it before the oyster numbers returned to their 1914 numbers?

d) Explain what happened in Malpeque Bay in terms of natural selection.

5 Charles Darwin first developed the idea of natural selection as the creative force behind evolution. The finches he studied on the Galápagos Islands were one of the organisms that led him towards this idea.

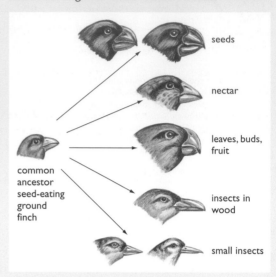

Large ground finches (the original type from the mainland) are coastal and use their short, straight beaks for crushing seeds	
Lowland-dwelling cactus ground finches use long, slightly curved beaks with a split tongue to feed on nectar of prickly pear cactus	
Vegetarian tree finches eat soft fruit/buds in the forest using their curved, parrot-like beaks	
Warbler finches live in the forest, catching insects in flight with their slender beaks	
Woodpecker finches live in forests. They have large, straight beaks and eat insect larvae	

On the mainland there is only one type of finch. Yet on the tiny Galápagos Islands there are six main types of finch and 13 separate species. They live on different islands or even in different parts of the same island. Use the theory of natural selection to explain carefully how this variety might have come about, based on change and adaptation within the original population of finches on each island.

Chapter 20: Selective breeding

- People use the principles of natural selection in the selective breeding of animals and plants for desired characteristics.

- Plants can be cloned to produce large numbers of genetically identical plants with desirable features.

- Micropropagation (tissue culture) is a way in which very small pieces of plants can be grown to produce commercial quantities of identical plants (clones) with desirable characteristics.

- Scientists have produced cloned mammals such as Dolly the sheep using the DNA from the cell from a mature organism and an egg.

- It is now possible to clone animals that have been genetically engineered to produce human proteins. There may be other ways of using them in the future.

Selective breeding

People have carried out selective breeding for centuries. They choose animals or plants with a desirable characteristic, such as a good milk yield for a cow or a big seed yield in wheat. They then breed from these organisms so that they pass on those desirable alleles to their offspring. Many different types of brassicas have resulted from an original wild plant. Selective breeding in plants is used to give advantages such as resistance to disease, heavy cropping, and the ability to survive in difficult conditions. Selective breeding in animals has been used to give characteristics such as heavier yields of milk, meat or eggs, producing more offspring, and resistance to disease and parasites.

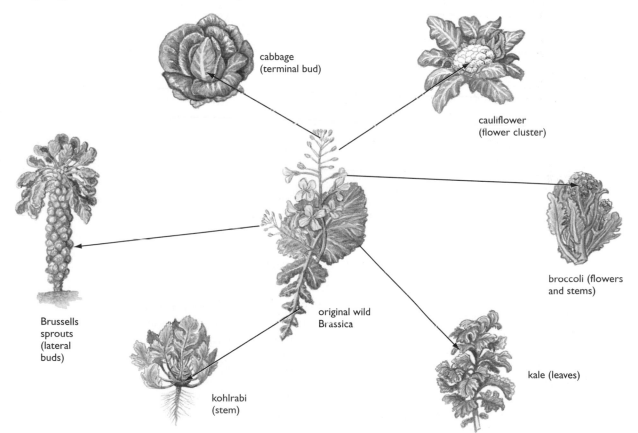

cabbage (terminal bud)

cauliflower (flower cluster)

broccoli (flowers and stems)

kale (leaves)

original wild Brassica

Brussells sprouts (lateral buds)

kohlrabi (stem)

Figure 20.1 *The effect of selective breeding in the brassica family*

Cloning plants

If you have a plant with particularly desirable characteristics, you can produce many identical copies (**clones**) of that same plant by **cloning**. Taking cuttings is the traditional method of cloning plants, but there is a limit to how many plants you can make using this method. The cuttings are kept damp using plastic films or special propagators.

Micropropagation involves taking tiny amounts of tissue from a plant (called explants) that are grown in special nutrient media. The plant tissue is supplied with hormones and all the minerals it needs to produce huge numbers of tiny new shoots. These are transferred to a different medium with hormones that cause the growth of roots. The tiny plants are then transferred to compost in a greenhouse until they become established plants.

There are many advantages to micropropagation. For example, large numbers of genetically identical plants can be produced rapidly, species that are hard to grow in other ways can be propagated, genetic modifications can be made in a small number of plants which then give thousands of plants carrying the desired change, tiny plants can be stored until needed and plants can be produced at any time of year. There are some disadvantages too. Because all plants are genetically identical they could all be vulnerable to a new disease or a change in conditions.

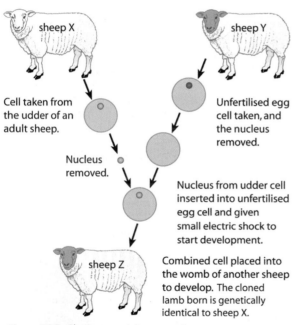

Figure 20.2 *Cloning an adult mammal*

Cell taken from the udder of an adult sheep.

Nucleus removed.

Unfertilised egg cell taken, and the nucleus removed.

Nucleus from udder cell inserted into unfertilised egg cell and given small electric shock to start development.

Combined cell placed into the womb of another sheep to develop. The cloned lamb born is genetically identical to sheep X.

sheep X

sheep Y

sheep Z

EXAMINER'S TIP ✔

Make sure you are clear about the differences between selective breeding, cloning plants, cloning animals and genetic modification.

Cloning mammals

Cloning animals is much more difficult than cloning plants, and cloning mammals is the most difficult of all. The best-known cloned mammal was the first, Dolly the sheep. Dolly was created by taking the diploid nucleus from the udder cell of a mature sheep and placing it in the egg cell of another sheep from which the nucleus had been removed. This was given a tiny electric shock, which started it dividing. The cell developed into an early embryo. It was then transferred into the uterus of a foster mother where the embryo grew and developed into Dolly. Cloning mammals has not proved easy. There are many failures and problems for every successful clone produced.

Sheep and other mammals can be genetically engineered to produce proteins that humans need to treat diseases. The genetically modified animals secrete the desired protein in their milk. Once one animal has been genetically modified, it can be cloned to produce a number of identical animals, all of which will produce the desired protein in their milk. Scientists hope that one day genetically modified mammals might be used to develop organs that could be used in human transplants without being rejected by the human immune system.

1 Copy and complete these sentences. Use the words below to fill in the gaps.

| plastic bags | characteristics | identical |
| damp | cuttings | |

New plants can be produced quickly and cheaply by taking _____ from older plants. These new plants are genetically _____ to their parents. This means that if cuttings are taken from a plant that has the _____ you want, all the cuttings will have them as well. Cuttings are most likely to grow successfully if they are grown in _____ conditions until their roots develop, so they are covered with _____ or sheets or put in special propagators.

2 We develop new varieties of animals and plants by choosing organisms that have useful characteristics and breeding from them. The animal or plant may end up looking very different from the original parent. For example, the wheat we use for flour has been selectively bred over thousands of years from wild grasses.

a) What do we call breeding animals and plants to get the characteristics we want?

b) Copy and complete this table of animals and plants that have been artificially selected for particular reasons

Animal selected	Reason why	Plant selected	Reason why
hens from wild chickens		wheat from wild grasses	large 'ears' for food
pigs from wild boar		potatoes from wild potatoes	
	large milk production, lots of meat		larger, sweeter fruit
dogs from wolves		garden roses from wild roses	

3 Large numbers of identical plants can be grown using micropropagation techniques.

a) How does tissue culture differ from traditional methods of taking cuttings?

b) Give four advantages of micropropagation over traditional methods of propagation by cuttings.

c) Although there are many advantages to these modern techniques, there is also a great disadvantage, which is that plant species propagated like this may not survive any major changes in conditions. Why is this?

4 *a)* What is happening to the number of people in the world?

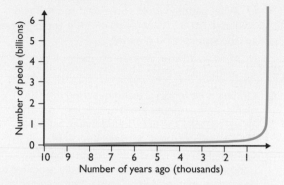

b) Why is it so important to increase the yields of many crops?

c) Many scientists are working hard to produce plants that not only give higher crop yields but also show resistance to disease. Produce a bar chart of the data in this table, which shows the percentage of the total crops lost around the world to disease and pests.

Crop	Percentage loss in 1987
wheat	9.1
maize	9.4
potatoes	21.8
vegetables	10.1
citrus fruits	16.4

Data from Central Statistical Office

d) Which crops were most affected by disease?

e) Why is resistance to disease as important as increased yields of crops?

5 Here is some information about the milk yields of three breeds of cattle over 60 years.

	Milk yield (kg/year)		
Year	Dairy Shorthorns	British Friesians	American Holsteins
1920	2000	2300	2300
1950	3000	3400	3400
1980	3800	5000	5800

a) Plot line graphs of these results.

b) Dairy Shorthorns are rarely seen nowadays except on rare breed farms. Why do you think this is?

c) Artificial selection has led to cows that can produce more and more milk. What do you think is the advantage of this?

d) Suggest any disadvantages.

e) What pattern would you expect to see if you looked at the meat yields of cattle and pigs over the same period of time?

f) How could you increase the meat yield of an animal like a pig?

6 Cloning is no longer a technique used only on plants. Animal clones are now possible too. The arrival of Dolly, the first cloned sheep, caused a considerable stir, and the technique continues to be used. Many of the sheep that are cloned are transgenic – they have been genetically modified to produce human proteins. Draw a flow diagram to show the stages in cloning a sheep.

7 Micropropagation techniques mean that, for example, 50 000 new raspberry plants can be gained from one old one. Using the old techniques, only two or three new plants would result. Extracting and cloning embryos from the best bred cows means that they can be genetically responsible for hundreds of calves each year instead of only two or three at most.

a) What are the similarities between cloning plants and cloning animals in this way?

b) What are the differences between the techniques for cloning animals and plants?

c) Why is there so much interest in finding different ways to make the breeding of farm animals and plants increasingly efficient?

Notes

Chapter 21: Using microorganisms

- People use microorganisms in a wide variety of ways.

- Yeast is widely used in the production of beer.

- Yeast produces carbon dioxide as a waste product of respiration.

- Bacteria known as *Lactobacillus* are involved in the production of yoghurt.

- Many useful microorganisms are grown on an industrial scale in fermenters, which provide suitable conditions for them to grow as quickly as possible.

Microorganisms

Most microorganisms are so small that they cannot be seen with the naked eye. They include protozoa, fungi, bacteria and viruses. Many microorganisms are vital as decomposers, some cause diseases, and some are used in the manufacture of useful chemicals and products. Fungi are used as food, to make bread, beer and wine, and to flavour cheeses.

Bacteria are used for many different things, including making yoghurt. They are also used in a genetically modified form to make human proteins such as human insulin.

Viruses can now be used in the process of genetic modification.

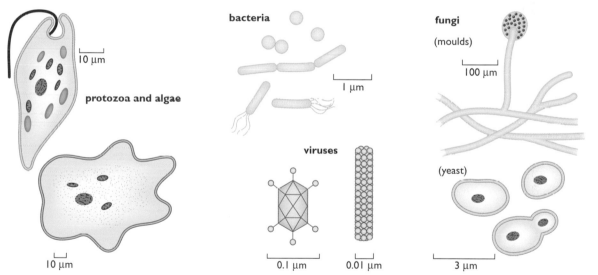

Figure 21.1 *Different types of microorganisms*

Fermentation and biotechnology

Biotechnology is the use of living organisms to make useful chemicals and products or to perform an industrial task. One of the oldest forms of biotechnology is the use of the **fermentation** reaction in yeast to make beer and other alcoholic drinks. When yeast respires anaerobically it produces ethanol, the chemical which acts as a drug in alcoholic drinks. The carbon dioxide produced by yeast is also used to make bread rise.

Glucose $\rightarrow$ ethanol + carbon dioxide

EXAMINER'S TIP ✔

The reaction of yeast with sugar is just one example of a fermentation reaction, not the only one! Many microorganisms undergo fermentation reactions of different types – this is why they are grown in fermenters.

Beer is made from barley. The process of making beer has not changed much over many years. It can be summarised as shown in the flow diagram below.

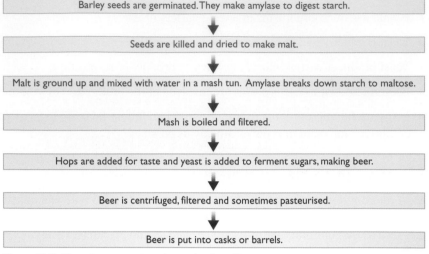

Figure 21.2 *Flow diagram to show the production of beer*

Other fermentation reactions involve bacteria. The production of yoghurt involves the use of special *Lactobacillus* bacteria, as shown in the flow diagram.

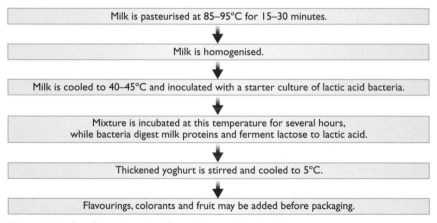

Figure 21.3 *Flow diagram to show the production of yoghurt*

Modern biotechnology

Microorganisms are now being used to make a huge range of food products and medicines, enzymes and even fuels. Many of these reactions take place in large vessels called fermenters. These are huge containers that hold up to $200\,000\,dm^3$ of liquid. They make it possible to control the environmental conditions such as temperature, oxygen and carbon dioxide concentrations, pH and nutrient levels, so that the microorganisms can grow and respire without being limited and can work as efficiently as possible. It is very important that everything in the fermenter is sterile, so that only the microorganisms that are wanted grow in the culture. Penicillin, a widely used antibiotic, is made in an industrial fermenter using *Penicillium* mould.

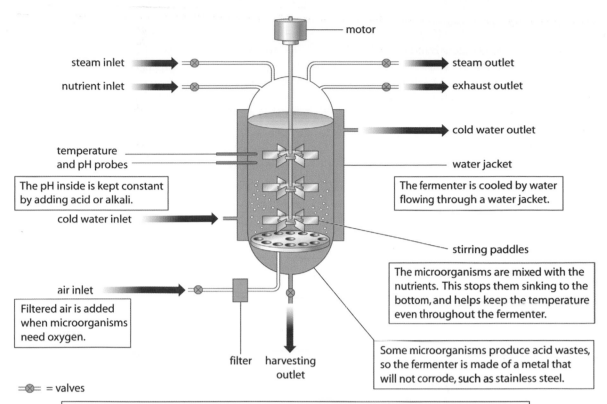

motor

steam inlet → ⊗ ⊗ → steam outlet

nutrient inlet → ⊗ ⊗ → exhaust outlet

→ cold water outlet

temperature and pH probes

The pH inside is kept constant by adding acid or alkali.

water jacket

The fermenter is cooled by water flowing through a water jacket.

cold water inlet →

stirring paddles

air inlet →

Filtered air is added when microorganisms need oxygen.

The microorganisms are mixed with the nutrients. This stops them sinking to the bottom, and helps keep the temperature even throughout the fermenter.

filter   harvesting outlet

Some microorganisms produce acid wastes, so the fermenter is made of a metal that will not corrode, such as stainless steel.

=⊗= = valves

The contents of the fermenter are monitored by special probes. These record the concentration of nutrients, temperature, pH, oxygen and carbon dioxide levels. The data is fed into a computer, which automatically controls the conditions in the fermenter.

Figure 21.4 *A fermenter used to grow microorganisms on an industrial scale*

Experimental evidence

You can demonstrate experimentally that anaerobic respiration takes place in yeast and that carbon dioxide is produced during the reaction. A sugar solution is made up with boiled water that contains no air, and yeast is added. A layer of liquid paraffin is poured on top to make sure that no oxygen-containing air gets in. Any gas produced is passed through limewater. This acts as an indicator – it is a clear liquid that turns cloudy when carbon dioxide bubbles through it. You should also be able to smell the ethanol produced by the yeast as it carries out a fermentation reaction. You need a control containing yeast that has been killed by boiling.

Questions

1 Copy and complete these sentences. Use the words below to fill in the gaps.

| fermentation | penicillin | microorganisms |
| biotechnology | enzymes | |

Many of the processes carried out by _____ are known as _____ reactions. They are often used in _____ to produce substances useful to people. These include _____, alcoholic drinks and medicines such as _____.

2 a) What are the main types of microorganisms?

  b) Which types of microorganisms are most commonly used by people?

  c) List as many things made using microorganisms as you can think of.

3 Write word equations to show the difference between aerobic and anaerobic respiration in yeast.

4 a) What is beer?

  b) Describe the process of making beer, including the role of yeast.

5 **a)** Explain why temperature control is vital for successful beer and wine making.

b) Sometimes a small amount of yeast is left in a bottle of wine to make sparkling wine or champagne.

i) Explain why the yeast is left in the bottle.

ii) What is the gas that makes the bubbles in the drink?

6 **a)** What is the main difference between the production of alcoholic drinks and the production of yoghurt?

b) Produce a flow diagram to summarise the production of yoghurt from milk *without* referring to page 86.

7 **a)** Give three reasons why microorganisms are so useful in industrial processes.

b) This diagram shows a fermenter that is used for growing particular bacteria in an industrial process.

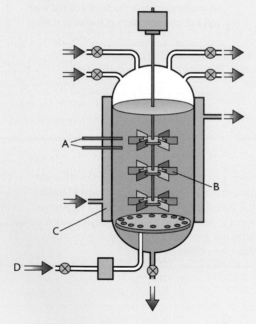

i) Name the parts labelled A, B, C and D.

ii) Explain the importance of the parts labelled A, B, C and D in ensuring that the best possible conditions for bacterial growth are maintained.

8 Why do the following factors tend to change during the fermentation process?

a) Temperature

b) Oxygen

c) pH

9 **a)** Copy this diagram and complete the labelling.

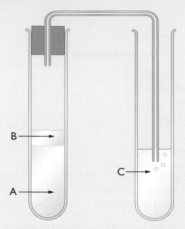

b) Explain how this apparatus could be used to investigate how temperature affects the respiration rate of yeast.

Chapter 22: Genetic modification

- DNA is the molecule of inheritance. It is made up of two strands coiled into a double helix. The strands are linked by a series of paired bases.

- A gene is a section of DNA.

- Genetic modification involves cutting a gene out of one organism and inserting it into the DNA of another. This is done using special enzymes to cut the desired gene out of one chromosome and stick it into another.

- Viruses and bacterial plasmids are used to transfer pieces of DNA from one organism to another.

- Genetically modified bacteria can be used to produce large quantities of human proteins such as human insulin.

- Plants can be genetically modified to improve food production in many ways, including improving their resistance to pests.

- A transgenic organism is one that has had genetic material transferred into it from another species, e.g. transgenic bacteria and transgenic sheep that produce human proteins.

DNA – the genetic material

DNA is the genetic material. It has a double helix structure held together by a series of paired bases. Adenine (A) is always paired with thymine (T), and guanine (G) is always paired with cytosine (C). The DNA molecule can replicate (reproduce) itself, which is why it is so effective as the genetic material.

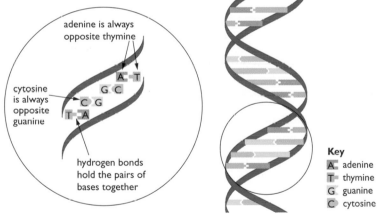

Key
A adenine
T thymine
G guanine
C cytosine

A sequence of three bases codes for an amino acid. A gene is a section of DNA that codes for a protein. Different genes code for different proteins because each has a unique sequence of DNA bases. The proteins produced as a result of the DNA code may act as enzymes to control the reactions of the cells or digestive system. They may be structural proteins, hormones, or proteins with specific jobs in the body such as haemoglobin or antibodies.

Figure 22.1 *The structure of DNA*

Recombinant DNA

In genetic modification (also known as genetic engineering), a gene from one organism is cut out and added to the DNA of another organism, often of a different species. The DNA from two different organisms combined that results from genetic modification is known as **recombinant DNA**. An organism that receives a new gene from a different species is known as a **transgenic** organism, or sometimes as a **genetically modified (GM) organism**.

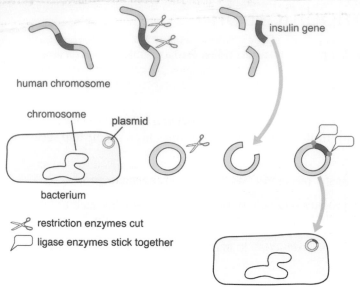

insulin gene

human chromosome

chromosome

plasmid

bacterium

restriction enzymes cut

ligase enzymes stick together

Figure 22.2 *Genetic engineering*

The transgenic organism will manufacture the protein coded for by its new gene, so it will be able to make a new and different protein, e.g. transgenic bacteria that have had a human insulin gene added to their DNA can make human insulin, which can be used to treat people with diabetes.

The enzymes used to cut out the gene from one organism are called **restriction enzymes**. The enzymes used to stick the gene into the DNA of another organism are called **ligase enzymes**.

Genetically modified bacteria

The first genetically modified organisms were bacteria, because they are single cells and it is relatively easy to see the effect of the inserted gene. Genetically modified bacteria are now used in a number of ways, such as in making human insulin, making enzymes for washing powders that work at relatively high temperatures, making human growth hormone, making bovine somatotrophin (cattle growth hormone), making human vaccines, etc.

Genetically modified plants

To produce a GM plant, you need to introduce a new gene into a plant cell and then produce a whole plant from the single modified cell. The technique is based on the bacterium *Agrobacterium* which inserts plasmids into plant cells. New genes can also be inserted using tiny gold pellets. The genetically modified cells are turned into whole new genetically modified plants using micropropagation techniques.

Plants have been modified to have many different characteristics: to give fruit that stays ripe for longer before it goes bad; to be resistant to weedkillers, frost, pests and diseases; to contain more vitamins or other nutrients than ordinary crops; to produce human antibodies; to produce human antigens to act as vaccines; to tolerate drought, high and low temperatures and high salt levels; to produce biodegradable plastics, and so on.

There are problems with some of these modifications. Pests may become resistant to the very chemicals in the plants that make them pest resistant, and diseases may become resistant too. There are issues about the use of nutritionally enriched crops in developing countries where they are most needed – they are expensive and often farmers need to buy them every year, rather than save some of the seed from the harvest. Some people are concerned about the consequences of genetically modified organisms interbreeding with wild organisms.

Questions

1 Copy and complete these sentences. Use the words below to fill in the gaps.

genetically modified	proteins
DNA	gene	genetic

Genes are made up of short stretches of _____. In a _____ organism a _____ from one organism is removed and added to the _____ material of another organism. The GM organism can make _____ directed by the new gene.

2 **a)** What do the initials DNA stand for?

b) What is a gene?

c) The bases in DNA are adenine, guanine, thymine and cytosine. How do these bases pair up in the complementary strands of DNA?

3 Human growth is usually controlled by the pituitary gland in the brain. If the pituitary gland does not make enough hormone, a child does not grow properly and remains very small. This condition affects 1 in every 5000 children. Until recently the only way to overcome this condition was to extract growth hormone from the pituitary glands of dead bodies, but it took many bodies to produce enough hormone to enable one child to grow properly. Genetic engineering means that pure growth hormone can now be produced in relatively large amounts by bacteria.

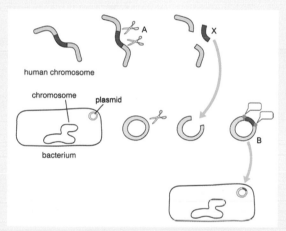

a) Copy and label this diagram. Use it to explain how a healthy human gene for making growth hormone can be taken from a human chromosome and inserted into a working bacterial cell to make it produce human growth hormone.

b) What are the advantages of producing substances such as growth hormone using genetic modification?

4 **a)** What is the job of insulin in the body?

b) Some diabetics can manage their diabetes by careful control of their diet, but about 250 000 people need to inject themselves regularly with insulin several times a day. From 1922 insulin from pig and cattle pancreases collected from slaughterhouses was used to help diabetics manage their condition. There were several problems with this as a source of insulin. What do you think they might have been?

c) Genetic engineering has made it possible for microbes to produce human insulin. In what ways do you think human insulin from microbes is better for people with diabetes than insulin from animals?

5 The process of genetic modification in plants differs from that in bacteria and animals. Using this diagram to help you, explain carefully the process of genetically modifying plants using *Agrobacterium*. Describe the role of restriction enzymes and ligase enzymes in the process.

6 In recent years there have been many changes in the way we can produce food and drugs, and treat diseases. Much of this biotechnology is of great benefit to all of us, but there are issues we need to think about. You are given a selection of comments below. Use them to help answer the questions.

1 Transgenic animals are the most recent success story of genetic engineering. Here a human gene for the production of a useful protein is engineered into an animal such as a sheep or a cow. When that animal lactates (produces milk), the human protein will be present in the milk and can easily be extracted and purified. This is already happening – there is a flock of sheep all producing a human blood clotting factor in their milk that can be supplied to people whose blood does not clot properly, without them running the health risk of getting the clotting factor from the blood of other people.

2 'My son is really short, but there's nothing wrong with him. Tall people always seem to get more respect and better jobs. I want my son to be given some of this new growth hormone stuff they're getting from bacteria, so he can end up 6 feet tall instead of only 5 feet 5 inches.'

3 Scientists are using genetic engineering techniques to breed pigs with special human proteins on their hearts and kidneys, so that their organs can be used for human transplants without any fear of the new heart or kidney being rejected. Some people find this idea objectionable, or worry about the risk of transferring diseases across the species, but doctors say that donor organs from people are in such short supply that they have to look at other possibilities.

4 Now that growth hormone can be produced in relatively large amounts using genetically engineered bacteria, there is a growing 'black market' for the drug as athletes and body builders try to buy it. In adults it seems that the hormone helps to increase the growth of muscle, and as one athlete said, 'Growth hormone occurs naturally in your body anyway, so it will be really difficult for the dope testers to pick it up.' It could be the performance-enhancing drug of the future.

a) Using the information given here and any ideas of your own, list some of the advantages to be obtained from the advances in genetic engineering and biotechnology.

b) Growth hormone is produced to help children whose bodies do not naturally make the hormone.

  i) Why do you think that parents with short or even average-height children might be prepared to pay for growth hormone treatment?

  ii) Why do you think athletes want to buy the hormone?

  iii) Do you think that either of these uses should be allowed? Give reasons for your answer.

7 Most people are happy to accept the benefits of using engineered bacteria to produce human proteins. Some people have concerns about putting human genes into animals such as cows, pigs and sheep. More people are unhappy at the idea of transplanting organs from an animal such as a pig into a person, even if they are genetically 'human'.

a) Why do you think people are less concerned about products from bacteria than those formed using sheep and cows?

b) What do you think about using organs from other animals in human transplants? Can you think of any other way the shortage of suitable donor organs might be dealt with?

c) Discuss the introduction and use of new biotechnology, including genetic modification of bacteria, plants and animals, giving examples of views both for and against its use.

Notes

Preparing for Exams

Getting the best possible mark in your exams is not just about thorough revision and learning your work. You need to approach the exams in the right way.

First of all, and most important, ***read the questions carefully***. Many students get lower grades just because they did not read the questions carefully, or missed out part of a question by mistake. Make sure that you read every question through several times before you answer it, so you do not lose *any* marks through carelessness!

Exam language

The language used in exams is very specific.

If an examiner asks you to name an organ, he does not want you to explain what it does. If an examiner asks you to explain something, she will not give you full marks if you only describe it.

So look carefully at what you are being asked to do at each stage of a question. Here are some explanations of exam terms to help you give the right answer.

Name

Do exactly what it tells you. Name something.

Example:	Name three cell structures.
Answer:	any three from nucleus, cytoplasm, cell membrane, cell wall (if a plant, bacterial or fungal cell), chloroplast, vacuole, mitochondria

Complete

Complete something that has already been started. For example, complete sentences by choosing words from a list, or complete a set of instructions for a practical investigation, or finish labelling a diagram from a list of terms or from memory.

Example:	Copy and complete the following sentences. Use the words below to fill in the gaps.
	carbohydrates malnutrition healthy fats energy chemicals
	It is important to eat the right amounts of food to remain fit and _____. Food provides you with _____ and the different _____ you need to keep your body working properly. The main types of food are _____, proteins and _____. Eating too much, too little or the wrong sort of food can result in _____.
Answer:	It is important to eat the right amounts of food to remain fit and **healthy**. Food provides you with **energy** and the different **chemicals** you need to keep your body working properly. The main types of food are **carbohydrates / fats**, proteins and **fats / carbohydrates**. Eating too much, too little or the wrong sort of food can result in **malnutrition**.

Describe

Exam questions often ask you to describe something. Examples of this type of question are as follows.

- Describe the structure of something. Sometimes the best way to do this is by using a labelled diagram.

Example: Describe the structure of the heart.

Answer:

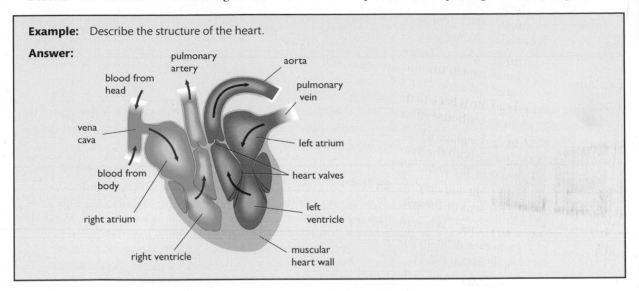

- Describe how you would do something.

Example: Describe how you would test food for the presence of starch.

Answer: Add a few drops of reddish-brown iodine solution to a sample of food. It will turn blue-black in the presence of starch.

- Describe how something works.

Example: Describe how the heart pumps blood around your body.

Answer: Deoxygenated blood flows into the right atrium from the body through the vena cava. At the same time oxygenated blood flows into the left atrium from the pulmonary vein. The atria contract to force blood through the tricuspid and bicuspid valves into the right and left ventricles. The valves close to stop blood flowing the wrong way. The ventricles fill with blood and then contract to force blood through the semilunar valves out of the heart. The right ventricle pumps deoxygenated blood out into the pulmonary artery to the lungs. The left ventricle pumps oxygenated blood out of the aorta to go around the body.

- Describe the advantages and disadvantages of something.

Example: Describe the advantages and disadvantages of sexual reproduction.

Answer: Advantages: introduces variety, makes survival more likely when the environment changes. Disadvantages: less certain, as it involves the meeting of two gametes and there is always a risk that the gametes will not get together or will not combine.

Give

This is usually used in the context of 'Give an example of …' Do exactly as it asks.

> **Example:** Give an example of a wind-pollinated flower and an insect-pollinated flower
>
> **Answer:** Wind – maize (or any other wind-pollinated flower). Insect – apple (or any other insect-pollinated flower).

Explain

Explaining something generally means not only describing it but also giving reasons why it is like that. This is more difficult and often more detailed than simply describing it. Candidates often lose marks in exams because they describe something when they have been asked to explain it.

> **Example:** Explain briefly but clearly the role of the placenta during pregnancy.
>
> **Answer:** You would *describe* a placenta as having a large surface area, but *explain* that a placenta has a large surface area to allow as much diffusion as possible to take place across it. You would *describe* a higher concentration of food and oxygen in the mother's blood, but *explain* that this higher concentration promotes diffusion across the placenta because of the concentration gradient. You would go on to explain that the fetus gets rid of waste products such as carbon dioxide and urea across the placenta into the mother's blood down another concentration gradient. The placenta grows into the wall of the uterus, so it anchors the fetus to the uterus wall. You could also mention that the placenta produces progesterone, the hormone that is needed to maintain the pregnancy.
>
> Extra marks are often awarded for a diagram. In this example, you would be expected to refer to the fetus, *not* the embryo, since 'embryo' describes only the first few weeks of gestation, during which time the role of the placenta is fairly small.

Discuss

When examiners ask you to discuss something, they are looking for evidence that you recognise at least two different angles or sides to the same question. You must be able to put forward at least two points of view and justify those points of view with scientific evidence where possible.

> **Example:** Discuss the introduction and use of new biotechnology, including genetic modification of bacteria, plants and animals.
>
> **Answer:** Your answer must include different examples of new biotechnology. It must also show an awareness of why the introduction of biotechnology such as genetic modification brings benefits, e.g. greater crop yields, ability to make pure human drugs, etc., and some of the concerns, such as the ability of people in developing countries to afford the new crops, concerns about breeding with wild species, etc.

active transport   The movement of substances against a concentration gradient and/or across a cell membrane, using energy.

ADH (antidiuretic hormone)   The hormone which controls the selective reabsorbtion of water in the kidney.

aerobic respiration   The process by which food molecules are broken down using oxygen to release energy for the cells.

allele   A version of a particular gene.

anaerobic respiration   Cellular respiration in the absence of oxygen.

asexual reproduction   Reproduction which involves only one parent and produces offspring which are identical to their parents.

auxins   Plant hormones which are involved in controlling the phototropisms.

carbohydrates   Food group which includes the sugars and starches. They are important for providing energy for the cells.

cell   The basic unit of all living organisms.

cellular respiration   Respiration which takes place in the cytoplasm and mitochondria of the cells.

cellulose cell wall   The rigid cell wall which surrounds plant cells.

chlorophyll   The green pigment contained in the chloroplast which captures light energy from the sun.

chloroplasts   The plant organelles which contain chlorophyll. They are the site of photosynthesis.

clone   Offspring produced by asexual reproduction which are identical to their parent organism.

codominance   Two alleles which are both expressed in the phenotype of an organism.

consumers   Organisms which feed on other organisms eg primary consumers eat plants, secondary consumers eat herbivores.

cytoplasm   The water-based gel in which the organelles of all living cells are suspended.

decomposers   Microorganisms that break down waste products and dead bodies.

denatures   The breakdown of the structure of a protein molecule if the temperature gets too hot or the pH changes.

diastole   The stage of the cardiac cycle when the heart fills with blood.

differentiate   The process by which unspecialised cells become specialised for a particular function.

diffusion   The net movement of the particles of a gas or a solute from an area of high concentration to an area of low concentration down a concentration gradient.

diploid   Having two sets of chromosomes in the nucleus of the cells: one from the male parent and one from the female.

disperse   The spreading of seeds away from the parent plant.

dominant   A characteristic which will show up in the offspring even if only one of the alleles is inherited.

double circulation   The separate circulation of the blood from the heart to the lungs and then back to the heart again and from the heart to the body and back to the heart.

ecosystem   All of the animals and plants living in an area, along with the things that affect them, such as the soil and the weather. An ecosystem includes all the interactions between the many different types of living organisms and the non-living components of their home.

egestion   The removal of undigested food from a cell or from the body in the form of faeces.

endocrine glands   The glands which produce hormones and secrete them directly into the blood.

enzyme   A protein molecule which acts as a biological catalyst, speeding up the rate of a specific reaction without being used up or affected. Enzymes are sensitive to both temperature and pH.

eutrophication   When a lake or river becomes enriched with nutrients, e.g. from fertiliser applied to fields, excess plant growth is followed by decay. Microorganisms use up oxygen from the water so that other organisms can no longer survive.

fermentation   Another term for anaerobic respiration which is particularly used for microorganisms such as yeast. Glucose is partly broken down into ethanol or lactic acid with the release of a small amount of energy.

fertilisation   The joining of the haploid male and female gametes to form a new diploid individual.

gametes The haploid sex cells which contain only one set of chromosomes. Male gametes include sperm and pollen, female gametes include ova and ovules.

genetically modified (GM) organism Organisms which have had DNA from a different species or a different individual inserted in their genome.

genotype The genetic makeup of an organism concerning the alleles of a particular gene, e.g. TT, Tt or tt.

global warming An increase in the temperature at the surface of the Earth due to greenhouse gases in the atmosphere trapping infrared radiation from the surface.

glycogen The carbohydrate energy store found in the liver and muscles of animals. It can be converted back into glucose when energy is needed in the cells.

greenhouse gases Gases such as carbon dioxide and methane in the atmosphere which absorb infrared radiation from the surface of the Earth and radiate it back to the surface, contributing to the greenhouse effect.

habitat The place where an animal or plant lives, including both the living and the non-living aspects of the area.

haploid Having only one set of chromosomes in the nucleus of the cell.

heterozygous When the two alleles for a particular gene on a pair of chromosomes are different, e.g. Tt.

homozygous When the two alleles for a particular gene on a pair of chromosomes are the same, e.g. TT or tt.

hormones Chemical messages which are secreted by endocrine glands and carried around the body in the blood to the organs they affect.

hypothalamus The part of the brain which regulates many body functions, e.g. temperature, thirst, hunger and sleeping. It also controls the production of many hormones from the pituitary gland.

invertebrates Animals which do not have a backbone.

meiosis A two-stage process of cell division which reduces the number of chromosomes from the diploid parent cell to the haploid daughter cell. It is important in the formation of gametes.

mitochondria A cell organelle which is the site of aerobic respiration, so they produce most of the energy needed by the cell.

mitosis Asexual cell division, a single-stage process which results in identical daughter cells.

natural selection The process by which evolution happens. Organisms produce more offspring than the environment can support. Only those best adapted to their environment – the 'fittest' – will survive to breed and pass on their characteristics.

neurone The basic unit of the nervous system, they are cells which carry minute electrical impulses around the body.

nitrogen cycle The continuous natural process by which nitrogen is exchanged between living organisms and the environment.

nucleus The organelle in the cell which contains the genetic material.

organ A group of different tissues working together to carry out a particular function in the body.

organ system A group of organs working together to carry out a particular function in the body.

organelles Membrane-bound structures found in the cytoplasm of a cell which carry out particular functions in the cell.

osmoregulation The control and regulation of the levels of mineral ions and water in the cytoplasm of a cell or in the blood of an organism.

osmosis The net movement of water down a concentration gradient from an area of high concentration of water molecules to an area of low concentration of water molecules across a partially permeable membrane.

phenotype The physical characteristics of an organism with respect to one or more particular genes.

photosynthesis The process by which plants make food combining carbon dioxide and water to make glucose using light energy captured using chlorophyll.

phototropisms The response of a plant through growth to light shining from one side only.

plasmids The extra circles of DNA containing extra genes found in bacteria and used by scientists in genetic engineering.

population   A group of individuals of the same species living in the same habitat and breeding together.

positive tropism   A tropism (movement by growth) in which the plant moves **towards** a stimulus.

quadrat   A piece of equipment used to sample an area to investigate the size of a population of plants or slow-moving animals. Quadrats are usually square frames of wood or metal that are laid on the ground to outline a sample area.

recessive   A characteristic which shows up in the offspring only when both of the alleles are inherited.

recombinant DNA   DNA which has had a section of DNA from another species or organism inserted into it.

sex chromosomes   The chromosomes which determine the sex of the offspring. In humans XX is a female and XY is a male.

sexual reproduction   Reproduction which involves the joining of two special haploid sex cells or gametes to produce diploid offspring which are genetically different from their parents.

synapse   The gap between two neurones where the transmission of the impulse is chemical rather than electrical. Transmission across the synapse depends on neurotransmitters.

systole   The stage of the cardiac cycle when the heart contracts and forces the blood out around the body and to the lungs.

transgenic   An organism which contains DNA from another organism.

transpiration   The loss of water vapour from the surface of the leaves of plants. This occurs through the stomata when they are opened to allow the gaseous exchange needed for photosynthesis to take place.

turgor   The state of a plant cell when water has moved in by osmosis so that the cytoplasm is pressing hard against the cellulose cell wall and no more water can enter the cell.

umbilical cord   The structure that connects a fetus to the placenta. Food and oxygen travel from the mother to the developing fetus through the umbilical cord, and waste products such as carbon dioxide and urea pass in the other direction.

vaccination   Giving a dose of a weakened or dead pathogen (a vaccine) to stimulate the immune system to produce antibodies and develop immunity to a disease.

xylem   The transport tissue in plants which carries water and mineral ions up from the root through the stems to the leaves.

zygote   The new diploid individual formed when the haploid male and female gametes meet and fuse in sexual reproduction.

Index

Index